AF449366

The WBF Book Series: Volume 3

ISA-95 Implementation Experiences

The WBF Book Series: Volume 3

ISA-95 Implementation Experiences

By WBF

Edited by
William Hawkins
Dennis Brandl
Walt Boyes

Momentum Press, LLC, New York

ISA-95 Implementation Experiences
Copyright © Momentum Press®, LLC, 2011

All rights reserved. No part of this publication may be reproduced, stored in a retrieval system, or transmitted in any form or by any means—electronic, mechanical, photocopy, recording, or any other except for brief quotations, not to exceed 400 words, without the prior permission of the publisher

First published in 2011 by
Momentum Press®, LLC
222 East 46th Street, New York, N.Y. 10017
www.momentumpress.net

ISBN-13: 978-1-60650-215-0   (hard back, case bound)
ISBN-10: 1-60650-215-8   (hard back, case bound)

ISBN-13: 978-1-60650-217-4   (e-book)
ISBN-10: 1-60650-217-4   (e-book)

DOI forthcoming

Volume 3 in the WBF Book Series by WBF, edited by William Hawkins, Dennis Brandl, and Walt Boyes, published by Momentum Press®, LLC

Cover Design by Jonathan Pennell
Interior Design by Scribe, Inc. (www.scribenet.com)

First Edition: February 2011

10 9 8 7 6 5 4 3 2 1

Printed in Taiwan

Contents

Tables

The purpose of this series of books from WBF, The Organization for Production Technology, is to publish papers that were given at WBF conferences so that a wider audience may benefit from them.

The chapters in this series are based on projects that have used worldwide standards—especially ISA 88 and 95—to reduce product variability, increase production throughput, reduce operator errors, and simplify automation projects. In this series, you will find the best practices for design, implementation, and operation and the pitfalls to avoid. The chapters cover large and small projects in a wide variety of industries.

The chapters are a collection of many of the best papers presented at the North American and European WBF conferences. They are selected from hundreds of papers that have been presented since 2003. They contain information that is relevant to manufacturing companies that are trying to improve their productivity and remain competitive in the now highly competitive world markets. Companies that have applied these lessons have learned the value of training their technical staff in relevant ISA standards, and this series provides a valuable addition to that training.

The World Batch Forum was created in 1993 as a way to start the public education process for the ISA 88 batch control standard. The first forum was held in Phoenix, Arizona in March of 1994. The next few years saw growth and the ability to support the annual conference sessions with sponsors and fees.

The real benefit of these conference sessions was the opportunity to network and talk about or around problems shared by others. Papers presented at the conferences were reviewed for original technical content and lack of commercialism. Members could not leave without learning something new, possibly from a field thought to be unrelated to their work. This series is the opportunity for anyone unable to attend the conferences to participate in the information-sharing network and learn from the experiences of others.

ISA 88 was finally published in 1995 as *ISA-88.01-1995 Batch Control Part 1: Models and Terminology*. That same year, partially due to discussions at the WBF conference, ISA chartered ISA 95 to counter the idea that business people should be able to give commands to manufacturing equipment. The concern was that business people had no training in the safe operation of the equipment, so boardroom control of a plant's fuel oil valve was really not a good idea. There were enough CEOs smitten with the idea of "lights-out" factories to make a firewall between business and manufacturing necessary. At the time, there was a gap between business computers and the computers that had infiltrated manufacturing control systems. There was no standard for communication, so ISA 95 set out to fill that need.

As ISA 95 began to firm up, interest in ISA 88 began to wane. Batch control vendors made large investments in designing control systems that incorporated the models, terminology, and practices set forth in ISA 88.01 and were ready to move on. ISA 95 had the attention of vendors and users at high levels (project-funding levels), so the World Batch Forum began de-emphasizing batch control and emphasizing manufacturing automation capabilities in general. This was the beginning of the transformation of WBF into "The Organization for Production Technology." Production technology includes batch control.

The WBF logo included the letters "WBF" on a map of the world, and since this well-known image was trademarked, the organization dropped the small words "World Batch Forum" entirely from the logo after the 2004 conference in Europe. WBF is no longer an acronym. Conferences continued annually until the economic crash of 2008. There was no conference in 2009 because many companies, including WBF, were conserving their resources.

WBF remained active and solvent despite the recession, so a successful conference was held in 2010 using facilities at the University of Texas in Austin. Several papers spoke of the need for procedural control for continuous and discrete processes. The formation of a new ISA standards committee (ISA 106) to address this need was announced as well. Batch control is not normally associated with such processes, but ISA 88 has a large section on the design of procedural control. There is a need for a way to apply that knowledge to continuous and discrete processes, and some of those discussions will no doubt be held at WBF conferences, especially if the economy recovers. We would like to invite you to attend our conference and participate in those discussions.

WBF has always been an organization with an interest in production technologies beyond batch processing, even when it was officially "World Batch Forum." Over the years, as user interests changed, so has WBF. We have not lost our focus on batch; we have widened our view to include other related technologies such as procedural automation. We hope you will find these volumes useful and applicable to your needs, whatever type of process you have, and if you would like more information about WBF, we are only a simple click away at http://www.wbf.org.

William D. Wray, Chairman, WBF
Dennis L. Brandl, Program Chair, WBF
August 2010

Foreword by Walt Boyes

Many years ago, some dedicated visionaries realized that procedure-controlled automation would be able to codify and regularize the principles of batch processing. They set out on a journey that eventually arrived at the publication of the batch standard ISA 88 and the development of the manufacturing language standard ISA 95.

Many end users have benefited from the work of these visionaries, who founded not only the ISA 88 Standard Committee but also the WBF. WBF has been an unsung hero in the conversion of manufacturing- to standards-based systems.

Today, WBF continues as the voice of procedure-controlled automation in the process and hybrid and batch processing industries. The chapters that make up this book series provide a clear indication of the power and knowledge of the members of WBF.

I have been proud to be associated with this group of visionaries for many years. *Control* magazine and ControlGlobal.com are and will continue to be supporters of WBF and its aims and activities.

I would like to invite you to come and participate in WBF, both online and at the WBF conferences in North America and Europe that are held annually. You will be glad you did. You can get more information at http://www.wbf.org.

Walt Boyes, ISA Fellow
Editor in Chief
Control magazine and ControlGlobal.com

The chapters in this book are written by people who have implemented systems with the aid of various parts of the ISA-95 standard. There is very little untested theory here. The chapters are divided into three sections:

- Chapters 1 to 3 are concerned with ISA-95 in general.
- Chapters 4 to 13 discuss ISA-95 and Enterprise Resource Planning (ERP)–Manufacturing Execution System (MES) integration
- Chapters 14 to 20 are about ISA-95 and MES

Almost all the chapters refer to MES because the business systems (ERPs) do not yet speak the control system's language, starting with transactions. Some chapters refer to SAP, a multinational business software company. WBF normally discourages commercial references, but SAP was the 800-pound gorilla in the room and needed to be mentioned. SAP claims 102,500 customers according to its Web site (http://www.sap.com/usa/index.epx).

Charlie Gifford's excellent Chapter 1 is an overview of the problem of communication between the shop floor and the transactional business machines. To fix this, it introduces thirty-eight acronyms used in ten times as many places. To help avoid confusion, I've made an effort to expand all acronyms that were not expanded in the original paper.

Scott Sommers's Chapter 8 supplies a good introduction to the five Levels of ISA-95. Dr. Ted Williams of Purdue University developed the Purdue Reference Model for enterprise manufacturing in the eighties, which gave us the Levels that put MES in Level 3.

Manufacturing Operations Management (MOM) models are required to make sense of MES. The introduction to them can be found in Jean-Luc Delcuvellerie's Chapter 16.

Bianca Scholten's Chapter 17 has a great story about interaction with customers, leading to a complete outline for writing a User Requirements Specification (URS) document for any ISA-95 integration project. The outline is in the chapter's appendix and really deserves special notice.

WBF's Business To Manufacturing Markup Language (B2MML) is essential for communication among MES and ERP systems. Chapters 4, 7, 9, and 11 discuss the uses of B2MML and provide examples of code.

A Note on Style

The WBF series has been edited using the 15th edition of the *Chicago Manual of Style* (CMS). This is the reason that you will find that your favorite capitalized words are not capitalized in certain instances. As a community of people that read books and papers on manufacturing process automation and business-shopfloor integration, we are used to every other word being capitalized because it is the name of a concept or a program variable, schema, and so on. The capitalization of terms in specification documents only makes it worse.

The problem is mitigated by using acronyms that are defined once in each chapter. Some words and phrases that are capitalized are important concepts that should be differentiated from common usage; others really are proper names. The rest contain the desired information whether they are capitalized or not, and so the CMS prevails.

Historical Perspective

Two kinds of computer control systems have evolved since technology made them possible. The first types of systems that automated business procedures such as payroll, manufacturing profit and loss, human resources, and others were all derived from a monetary viewpoint. The second types of systems automated process control, beginning with the basic Proportional-Integral-Derivative (PID) loop and adding control functions as the computer and glass control panel became accepted.

Some business computer vendors tried to get into the process control field and failed for lack of knowledge of what they were getting into. Process control is concerned with maintaining and recording process measurements, not profit and loss. Operators want to know how the process is doing and want to have handles that will let them take corrective action, with less than a second elapsing between command and response.

And so it was that a great divide opened up between manufacturing control and business IT. Manufacturing requires uninterruptible computing power, with service required all day, every day of the week. Everyone in manufacturing has a story about a failure at 3 a.m. Business machines operated weekdays, and users didn't mind having a few days off while the system was maintained. "Oh, I'm sorry but the computer is down right now. Can you call back tomorrow?"

This situation began changing as computer operating systems and communication systems settled down into something resembling standard operation. Businesses knew they needed information from Manufacturing because they'd been getting paper status reports, perhaps weekly.

All this was in flux when ISA95 was born in 1995. The original impetus came from the extravagant claims of MES vendors saying they could connect the boardroom to the shop floor. It is not to their credit that they did not understand that people in upper management had no training in the safe operation of their manufacturing facilities. There was no control knob for "profit" that the CEO could crank to its extreme position.

There is still a lot of talk about the shop floor. When the day comes that there is nothing more to do on the shop floor, marketing will discover that there is a basement under the shop floor and that it has data. Don't ask what data—they'll figure it out as they go. That will set off another round of product differentiators and books like this one.

Distributed Control Systems (DCSs) use parallel processing to crunch incomprehensible amounts of data. PID controllers may save some data from the last one or two iterations, but they are mainly concerned with current measurements and previous outputs. Historians may be able to capture the results of each cycle, but this fire hose stream of data needs to be condensed before a human can make use of it. Before computers, data were recorded on strip charts that moved at 3/4 of an inch per hour, making anything less than a minute or two lost in the width of the ink trace.

ISA95 set out to model the communication between business systems and the Manufacturing Control Systems (MCSs). First you have to model what's out there on the shop floor, then decide what's worth remembering, then organize it by manufacturing function, and finally you need two-way communication using a formal language between manufacturing and business. You can't design a language without knowing what has to be communicated, hence the need to make data models of manufacturing and business activities.

Communication means secure transactions to business people. Control people are more used to the publish-and-subscribe model because there is not enough time for secure transactions. Subscribers have ways of knowing when publication stops, just as you know when your newspaper wasn't on the doorstep. See Chris Monchinski's excellent Chapter 4 for a description of the use of publish and subscribe in business, as well as using ISA-95 principles when a full MES layer is not justified.

One other aspect of communication is not discussed in these chapters, and that is time synchronization among systems. Some systems are islands in time, updated occasionally by somebody's wristwatch. An increasing number of systems use network time applications, like Simple Network Time Protocol (SNTP), to synchronize machines to Global Positioning System (GPS) time. GPS time receivers can be used to synchronize networks that are isolated by a data diode.

Considering all the messages that have to be time stamped, this is a good thing. The U.S. Food and Drug Administration (FDA) now requires time stamping as close to the source as possible. Foundation Fieldbus was designed to provide this capability in the early nineties. The limit of time resolution is set by the execution cycle of the control function blocks. A block must execute to get a value or announce an alarm, and that execution time is set by the macrocycle for block execution. Nobody executes all of them at once.

Defining a language requires syntax, vocabulary, and grammar that can be understood by both sides. Furthermore, computer networks require addresses and rules for packing a message into a bag of bits and also require ways to handle errors. Transmission Control Protocol (TCP) is a secure protocol because it uses handshake messages ("Did you get it?" "Yes, I got it") to establish communication for every bag of bits sent. TCP has rules for finding lost messages and recovering them by retries. The principle language for MES communication is now WBF's markup language B2MML, which is also used for Manufacturing To Business (M2B).

Markup languages began at IBM in the seventies with Generalized Markup Language (GML), invented by Goldfarb, Mosher, and Laurie. Later GML became Standard Generalized Markup Language (SGML). The first major application was Hypertext Markup Language (HTML), which allows zillions of computers to talk to each other using Web browsers. HTML evolved at a rapid pace, particularly as a result of conflicts between Netscape and Microsoft, and has now become "polluted" by all the stuff added for better marketing, tracking, and advertising.

Extensible Markup Language (XML) was privately developed by Bosak, Bray, Clark, and others in the nineties to restore purity. The World Wide Web Consortium (W3C) then developed Extensible Hypertext Markup Language (XHTML). Sun Microsystems introduced Java at about the same time, which was picked up by IBM and Oracle for communication among their many operating systems. Microsoft chose XML and wove it into the .NET system, mostly because it wasn't controlled by Sun. Now IBM has adopted XML, and so it goes.

Markup languages contain elements and attributes that both ends of the communication link must understand. An XML schema is a document that defines the elements and attributes, limits on the way they may be structured, their syntax, and the data types that may be used in a message. B2MML is a schema for the XML documents used for communication on a network of mixed MES and business system computers. Of course, XML is not used for any of the traffic on a control network because it is too verbose, but there may be interface or gateway computers that use it.

Eventually, the Wild West that was MES adopted ISA95's work to home in on a standard way of selecting manufacturing data and exchanging it with ERP systems. That work is still evolving, just as HTML is evolving to fit the needs of users of the World Wide Web.

Bill Hawkins
August 2010

CHAPTER 1

ISA-95 Business Case Evolves through Applications and Methodologies

Presented at the WBF
North American Conference,
March 5–8, 2006, by

Charlie Gifford
President, Chief Manufacturing Analyst
charlie.gifford@cox.com
21st Century Manufacturing Solutions LLC
630 Angela Drive
PO Box 4424
Hailey, ID 83333, USA

Paresh Dalwalla
President
pdalwalla@optebiz.com
OpteBiz Inc.
5333 South Main #B
Sylvania, OH 43560, USA

Abstract

Business To Manufacturing (B2M) data exchange applications and system life-cycle methods are being developed from the ANSI/ISA-95 Enterprise-Control System Integration Standard to adapt and optimize manufacturing in the 21st century "pull" marketplace. The MESA/ISA-95 Best Practices Working Group will publish an annual ISA Technical Report to document these evolving applications and methods with an explanation of the ISA-95 business case. The business case centers on the following:

- Lowering life-cycle cost of B2M interfaces and manufacturing operations applications

- Constructing the flexible Manufacturing Application Framework (MAF) to optimize B2M interoperability and production (capability) flexibility through B2M functional segregation for optimized production workflow

The standard practically addresses today's B2M language (terminology and schema) requirement for an Application To Framework (A2F) data exchange using Service-Oriented Architectures (SOA). Utilizing ISA-95, schema foundation, applications, methods, and business cases are established through a structured MAF consisting of (1) workflow function organization, (2) transformation best practices for operations applications, and (3) their transactional interfaces. The business case for ISA-95 is demonstrated by the fact that ISA-95 does the following:

- Enables application of Lean practices of Standard Work and value streaming

- Enables the development of B2M functional segregation methods to correctly position operational tasks within Enterprise Resource Planning (ERP), Supply Chain Management (SCM), and Manufacturing Execution Systems (MESs) to optimize single-piece workflow and supply chain flexibility

- Structures a life-cycle management framework to lower a system's Total Cost of Ownership (TCO) and execute MAF flexibility in response to market change

The MAF utilizes ISA-95 part 3, "Models of Manufacturing Operations Management," to explain the influence of different system configurations on workflow, life-cycle cost, flexibility, and change management.

Recognizing a Historical Inflection Point in World Industry and Markets

Beginning in 1995, the Instrumentation, Systems, and Automation Society (ISA) and WBF developed the ANSI/ISA-95 Enterprise-Control System Integration Standard and the Business To Manufacturing Markup Language (B2MML). These works are intended as the foundation for standardized best practices for information exchange between plant systems and plant-to-business systems. Over the last 10 years, Manufacturing Operations Management (MOM) solutions evolved to enable the distributed supply chain networks for 21st-century markets. ISA-95 based MOM applications and methods are recognized as the foundation for configurable, interoperable software tools to integrate interoperable data in readily useful forms to extended enterprise systems. This chapter assumes that the reader is familiar with ISA-95 and B2MML and so will focus on best practices and business cases and will not provide a standards overview.

Goal of the "ISA95/MESA Best Practices Technical Report"

The annual "ISA95/MESA Best Practices Technical Report" will explain how the ISA-95 Enterprise-Control System Integration Standard is applied to lower TCO of MOM systems and their enterprise and plant interfaces. ISA-95 best practices propose a three-legged MAF containing the following:

1. *Tools.* ISA-95 methods and technical applications that characterize, support, and adapt production workflow processes

2. *Training and staffing.* Defined system roles and skill sets for personnel for MOM processes

3. *Delivery.* A defined transformation and life-cycle management process for MOM

The ultimate goal of the ISA-95/MESA Best Practices Working Group is to explain "how to" apply, migrate to, and maintain a single data definition across Level 3 functions and interfaces, the MOM domain, and their Level 4 domain enterprise interfaces. By utilizing developing ISA-95 methodology and technical applications, the TCO for manufacturing IT architectures as well as manufacturing and supply chain operational costs are dramatically reduced. The annual "ISA-95 Best Practices Technical Report" consists of a series of related "how to" white papers described in the context of ISA-95 models, definitions, data flows, and the Level 3 interfaces between Level 4 enterprise functions and Level 2 shop floor systems.

Manufacturing Trends Relevant to the Role of ISA-95

The 21st-century manufacturing model is all about flexibility of production capabilities within globally distributed supply chain networks or Demand-Driven Supply Networks (DDSNs; as described by AMR Research). Manufacturing markets are rapidly changing and driven by global competitive trends that make production flexibility a critical path component of supply chain collaboration. This coordinated data exchange across global supply chains and internal enterprise groups is just a part of the ISA-95 business case. Current industry discussions are focused on production's actual role in e-commerce, product development, supply chain planning and replenishment, or logistics. For any 21st-century manufacturer to be competitive, actual manufacturing operations activities must be highly interactive in supply chain and enterprise processes for effective collaboration and competition. This is the domain of collaborative and flexible MOM system architectures. This chapter explains the business cases for using evolving ISA-95 methods to effectively design, implement, change, and optimize the MOM business processes and for supporting MOM system architectures within the larger DDSN model.

Global Business Drivers for Flexible Manufacturing

Each vertical industry is being influenced by their unique combination of the following global business drivers for flexible manufacturing:

- Increased globalization: global markets with distributed sources of supply, production, and distribution facilities
- Increased customer diversity: culturally and geographically
- Increased access to competitive data
- Increased level of expected value
- Increased outsourcing of production and logistics operations
- Increased pace of new product introductions
- Increased product quality at lower cost

Today's competitive environment requires new business models that accommodate changing geographic presence, cost base, product array, use of new materials and technologies, and relationships with customers, suppliers, and other trading partners (Fig. 1.1).

The other side of the challenge includes organizational issues and aligning the goals and objectives of the different players in the organization as the company designs their 21st-century business model. Manufacturers that adopt a standardized approach to understanding, implementing, and deriving benefits from MOM applications should be able to accelerate the transformation to a successful implementation while increasing satisfaction levels of users.

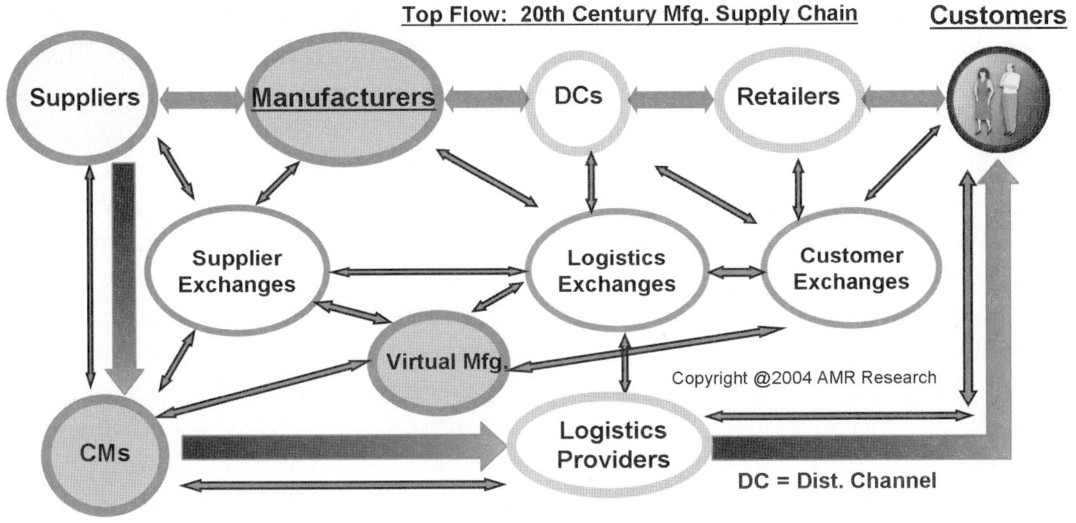

Figure 1.1. Aligning manufacturing capabilities to 21st-century challenges mandates change.

The B2M integration of operational systems requires unique skills to accelerate the B2M transformation. A thorough understanding of the following is essential:

- Operational processes and workflows
- Real-time controls and automation systems
- Information technology systems
- Knowledge of standards that encompass all levels of information flow
- Lean manufacturing and Six Sigma methods

The current MOM skills sets available in the marketplace are often segregated across these knowledge items by different levels; team members must become more broadly based to successfully integrate MOM architectures horizontally as well as vertically across the different levels.

DDSNs Create the Interactive 21st-Century Manufacturing Model

To effectively compete in 21st-century markets, companies are creating improved tangible value by accurately matching products and value-add services to each customer's need within hours rather than weeks. Timing is the key to profitability now. This value objective is not new, but maturing capabilities of Web technologies (integrated into application software) have provided the interactive tools for the required collaborative and interoperable communication. In conjunction, ISA-95 is defining MOM data structures and exchanges (definitions) that can be utilized to construct the MOM Web services and SOA for MOM solutions. The operations management data and transaction definition in ISA-95 parts 3 to 6 are the basis of SOA for Manufacturing (SOAM) or a Manufacturing Services Architecture (MSA). These ISA-95 parts are establishing the real-time basis to quantify cycle time, cost, and resource elements of workflow for production, maintenance, inventory, and quality. As seen in Figure 1.5, B2M data flows and metrics are defined in terms of four B2M categories of information (activity definition, activity capability, activity schedule, and activity performance) for the four primary plant activity models shown. These data elements are required for scheduling and planning order fulfillment across DDSNs. In 2005, ISA95 and the Supply Chain Council formed the ISA95/SCOR Alignment Working Group, which mapped data flows between the Supply Chain Operations Reference (SCOR) and ISA-95 part 3 models. This is the foundation for aligning the development of the two standards, as ISA-95 parts 4 and 6 are composed over the next few years. This combined work enables the rapid evolution of MSA and DDSN architectures.

SOA Components

Over the last 15 years, integration technology has evolved from data to process level capabilities with SOAs being the merger of business process management and the enterprise services bus. SOA surrounds Web services containing the business process rules with various technologies to manage, orchestrate, and choreograph Web services into an executable business model. Core SOA services include the following:

- Services registry
- Enterprise services bus
- Web services management
- Web services security and identity management
- Web services development and programming tools

The business justification for ISA-95 based MSA within interactive MOM solutions is reinforced by contract manufacturers. The necessary practice of outsourcing production to contract manufacturers is driven by global markets in growing economies such as India, China, and Eastern Europe, across all industries. The result is low-cost competition for North American and European suppliers, due to lower labor and operating costs. Also, outsourcing has accelerated adoption of the DDSNs model where consumer demand for a single product order is now met by evaluating competing supply chain paths. Basically, the order fulfillment path is now determined by evaluating real-time supply chain cost to customer demand for on-time delivery at a specific quality level. Order commitments are made based on this algorithm (Fig. 1.2).

Original Equipment Manufacturers (OEMs), such as IBM, GE, and HP, previously known for building a variety of products, are now known for their market-leading product designs and their ability to market and sell them by managing their DDSNs through contract-manufacturer partners and internal production. MSA-based MOM solutions provide the means for OEMs to identify available materials and resources (capacity) across competing supply chains in a lead time versus price form for immediate prototyping to market demand (i.e., Design For Supply [DFS]). The more real time the SCM, the larger the profit margin due to order accuracy.

As OEMs evolve their MSA practices, they build a tighter relationship with key suppliers and contract manufacturers to (1) provide forecasted demand from all customers and sales channels and (2) require real-time production records and visibility to the OEM customer from ISA-95 based MOM solutions (Fig. 1.3).

To address global competition, 21st-century manufacturers are rapidly adopting several types of corporate software systems to transform to their global business model: ERP, SCM, Supply Chain Execution (SCE), Customer Relationship

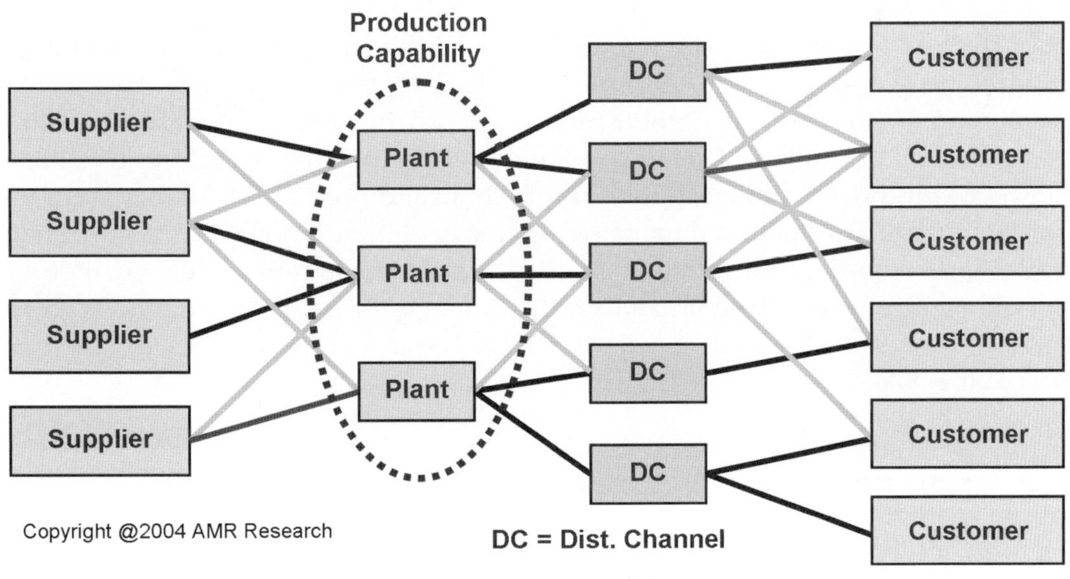

Figure 1.2. Business evolves into configurable DDSN.

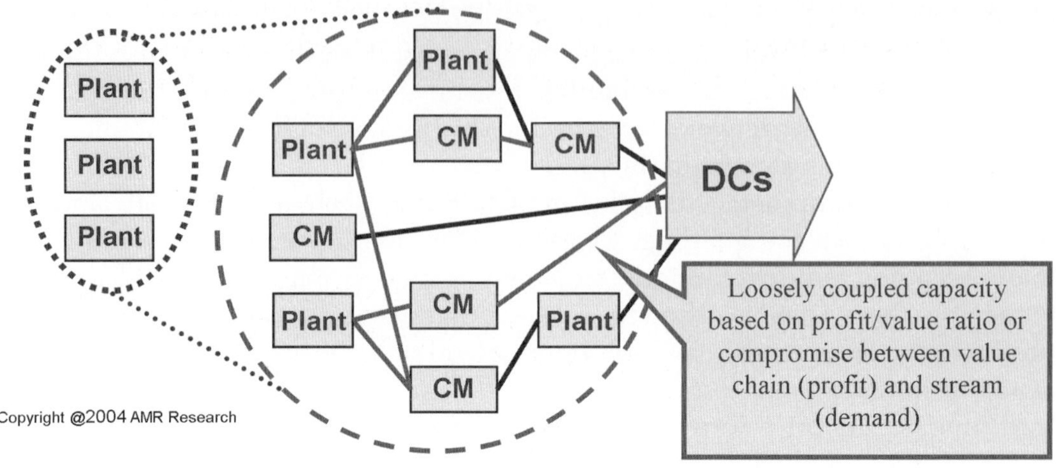

Figure 1.3. Production capability configured in real time: Evaluate customer value and On-Time Delivery (OTD) to production path and profit margin.

Management (CRM), design collaboration tools, Product Life-cycle Management (PLM), and others. These extended enterprise systems were supposed to be designed to exchange information outward to customers and suppliers in near real time. Results over the last 10 years have been poor to fair, since these tools were not designed for a DDSN global market. For some companies, these early generation systems provided quick benefit by reducing the time and costs of interacting with their 20th-century linear supply chain partners; however most have not achieved the predicted benefit due to lack of actual production data integrity (response and accuracy). An ISA-95 methodology for data exchange is being developed for the next-generation ERP system to support the complex (nonlinear) global DDSN model. This methodology will help map data and transactions between workflows of a distributed supply chain and production. With inaccurate or "too coarse" production data being the major limitation of the Y2K corporate business systems, a similar major issue is a lack of commonly defined metrics or schema across supply chain and production operations. ISA-95 methods solve these limitations for system interoperability. These will be further explained as part of the ISA-95 business justification.

ISA-95 Blends the SOA Approach into MOM

The ISA-95 standards address the interface or exchange of data between the extended enterprise systems (sales, planning, scheduling, and procurement) and the following part 3 models of MOM (and example MOM systems):

- *Production Management Operations*. Product tracking and tracing, manufacturing execution, manufacturing intelligence portals, finite capacity and detailed scheduling, work order management, production sequencing, batch execution, recipe management, and so on

- *Maintenance Management Operations*. Asset management; computerized maintenance management; preventive maintenance; Maintenance, Repair, and Operations (MRO); and so on

- *Quality Test Management Operations*. Statistical process control, statistical quality control, laboratory information management, Corrective And Preventive Action (CAPA), Material Review Board (MRB), and so on

- *Inventory Management Operations*. Management Resource Planning (MRP), tracking of plant-side raw material, Work In Process (WIP), finished goods, and so on

The following lists the ISA-95 foundation for MSA, that is or will be contained in parts 3 through 6:

- *ANSI/ISA-95.00.03-2005 Part 3: Models of MOM*
- *ISA-95.00.04 (Draft) Part 4: Object Models and Attributes of MOM*
- *ANSI/ISA-95.00.05-2007 Part 5: Business to Manufacturing Transactions*
- *ISA-95.00.06 (Proposed) Part 6: MOM Transactions*

With this work, the MSA concept brings the previously mentioned MOM operations of a manufacturing business into alignment to intelligently respond to market forces. Currently, corporations are attempting to identify MSA practices to publish and distribute customer demands across the supply chain. Once all suppliers in company's DDSN are able to align on market demand, products are then rapidly and accurately developed for new markets while maintaining high margins. The combination of a DDSN model being driven by an ISA-95 MAF allows a manufacturer to capture a large market share or even create markets, due to their ability to rapidly adapt their collaborative production resources to real-time market demand.

After Part 6, B2MML Is Required to Meet MSA Requirements and End-user Demand

B2MML version 4 is based on the following first approved, untested versions of ANSI/IEC/ISO/ISA-95 parts 1 and 2 in 2000 through 2002:

- 2000: *ANSI/ISA-95.00.01, Enterprise-Control System Integration, Part 1: Models and Terminology*

- 2002: *IEC/ISO 62264-1 Part 1: Models and Terminology*

- 2001: *ISA/ANSI-95.00.02 Part 2: Data Structures and Attributes*

- 2002: *IEC/ISO 62264-2 Part 2: Data Structures and Attributes*

In the current 5-year review of the first versions, many changes are being proposed due to end user lessons learned in applying the standards and schema applications (not to be addressed in this chapter). However, it is important to note that B2MML version 4 is a special case of manufacturing environments and the B2M interface instance, as opposed to a general case. This is illustrated in Figure 1.4 and Figure 1.5. B2MML version 4 is based on an academic definition at the Level 3 to Level 4 interface described in parts 1 and 2, where all MOM functions are plant side systems. B2MML schema has not yet evolved (and will not be able to evolve, since schema must follow the approved standard) to address a more wide range of real-world B2M interfaces where many MOM functions are within centralized corporate applications. With the release of part 3 and 5— and the eventual completion of parts 4 and 6—B2MML is evolving to adequately address Level 3 MOM data and workflows for a majority of hybrid manufacturing environments.

Most plants are hybrid environments from dock (raw materials) to dock (finished goods packaging) with a mix of work order types for customer orders (e.g., 80% Make To Stock [MTS], 10% Make To Order [MTO], 10% Engineer To Order [ETO]) and more of a mix of WIP work cell orders (50% MTS, 35% MTO, 15% ETO or rework). The B2M interface line ("B2M" line in Fig. 1.5) is determined by the MOM applications required at the plant floor to address the complexity of the work order mix and its associated workflow business rules. As evolving global markets drive manufacturers to rapidly adjust their work order mix based on market demand and drivers (i.e., toward a higher percentage of MTO and ETO), B2M functional segregation of MOM applications and the representative B2M interface line are determined by the MOM architecture that optimizes production, single-piece flow, profit margin, and throughput. Basically profit margin drives the MOM architecture. A manufacturer's ability to rapidly adapt their MOM architecture to new market conditions determines the level of success in their global markets. This proposed ISA-95 methodology is Step 3.5 in the MAF best practices methodology outlined later in the chapter.

The following is a list of the components of a simplified workflow complexity matrix:

1. Production types

   - Discrete manufacturing

   - Batch processing

   - Continuous processing

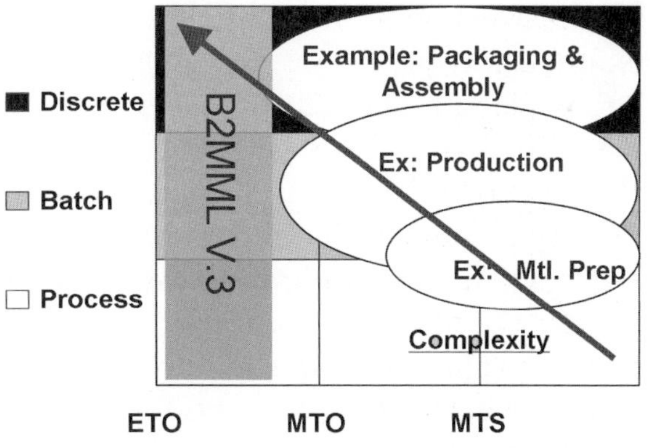

2. Work order types

   - ETO

   - MTO

   - MTS

   - Nine primary combinations with many hybrids (each have a specific set of business processes and rules)

   - Complexity contributors (e.g., product, legacy, speed, volume, color, size, compliance, SKU count)

Figure 1.4. Simplified workflow complexity matrix.

Once industry has agreed upon the standard form of B2MML, the ISA-95 body of work (schema, standards, applications, and methods) will adequately model the majority of the Level 3 MOM use cases, data flows, transactions, business processes, and metrics (interface, Key Performance Indicators [KPI], and operational) construction. Based on this MOM use case modeling, vendors will roll out their collaborative libraries of MSA Web services that end users essentially need for global DDSN architectures. At this point (2008–2010), the ISA-95 methodology

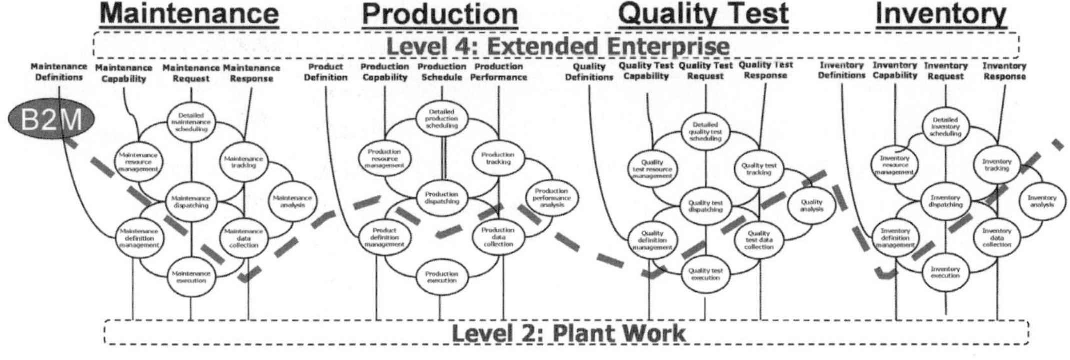

Figure 1.5. ISA-95 part 3 hypothetical B2M interface chart: Real-world B2M interfaces exist within MOM activity model functions.

for construction of the MAF will coalesce and mature into a proven SCM system required for the DDSN implementation. This chapter and the ISA-95 Technical Report (First Edition 2006) proposes the working methodologies to drive toward these goals over the next few years. Software vendors and MOM literature are moving to this important inflection point for the next 3 to 5 years to meet the 21st-century manufacturing model requirement. As of 2006, the ISA-95 body of work needs accelerating to limit the scale and variation of implemented beta B2MML interfaces and MOM applications. Many early innovative adopters are struggling with how to apply and extend B2MML version 4 to address their hybrid environments and are taking their best guess as to where the ISA-95 body of work will direct software vendors in the future.

Industry collaboration is the real-world challenge in order to lower TCO for integrated systems through focused, dedicated effort. Accordingly, the end user's commitment to B2M interoperability will remain speculative until this open standard's work has been organized and accomplished. Industry analysts, vendors, and end users are all looking to each other for leadership for this MOM standards effort. No clear leader has yet emerged as of this writing. A loosely coupled group of end users, vendors, and consultants are proactively forming alignment working groups, but the progress is much slower than the market need.

The ISA-95 Business Value: Low Cost B2M Interfaces and Flexible MAF

Manufacturing data exchange applications and system life-cycle methods are being developed from ISA-95 standards. ISA-95 applications and methods are being globally applied by innovative manufacturers to adapt and optimize manufacturing for the 21st-century "pull" markets.

The ISA-95 business case is centered on the following:

- Lowering the life-cycle cost of B2M interfaces
- Constructing the flexible MAF to optimize B2M interoperability and production (capability) flexibility through B2M functional segregation for optimized production workflow

The standard practically addresses the language (terminology and schema) requirement between operations and business systems for an Application To Framework (A2F) data exchange required for MSA.

The business case is further established by providing organization and best practices to transform operations applications and their transactional interfaces into a flexible manufacturing framework. Through its B2M functional and object models and single Extensible Markup Language (XML) schema hierarchy for B2M integration, ISA-95 transformation methods merge production workflows into the overall collaborative business process. The merger of ISA-95 models, applications, and methods to form the MAF is proposed by the ISA-95/MESA Best Practices Working Group in the annual ISA Technical Report. MAF defines data exchanges and metrics for integrating the production systems (1) horizontally between MOM applications and (2) vertically between global DDSN and enterprise systems.

ISA-95 business value and MAF are derived from two classes of best practices to illustrate the high value of the MOM application development and life-cycle processes utilizing ISA-95. Both classes of best practices are early in their life cycle and are rapidly advancing due to endorsement by the end user and vendor community such as SAP and Microsoft. These ISA-95 best practices are being applied in isolated MOM projects by the most innovative manufacturers in the world, such as Procter & Gamble, Dow, Arla Foods, Nestlé, DuPont, and BP. Widespread use has not occurred due to the immature state of MOM solutions and SOA manufacturing technologies.

Class 1: ISA-95 Technical Applications for Improving B2M Interface Interoperability

The ISA-95 technical applications utilize the following:

- B2MML schemas
- ISA-95 models from parts 1 to 3
- Proposed part 4 Level 3 MOM data flows and operations metrics
- Part 5 B2M transactions
- Proposed part 6 MOM transactions
- Other developing information technologies such as XML, Web services, SOAs, or data exchange frameworks for application interoperability

Class 2: ISA-95 Transformation Methodology for Improving B2M Operations Interoperability

Class 2 describes a set of operations transformation methodologies that align with other current best practices for the software development life cycle such as Good Automation Manufacturing Practices (GAMP) or Microsoft Enterprise

Framework (MEF) methods. These ISA-95 based methods are intended to be used to construct an MAF for optimizing B2M operations interoperability by providing MOM system architectures that are able to adapt to market changes in the company's DDSNs.

This chapter simply outlines and briefly describes the "ISA-95 Best Practices Classes." This white paper introduces the evolving best practices and explains the high-level business case through market need. In future white papers from the ISA95 Best Practices Workgroup, a business justification for each application and methods will be explained using the following five steps:

1. List the proposed ISA-95 best practice
2. Identify the prioritized business driver and operational benefits
3. Identify the current state and the underlying forces (reasons for underperformance, stakeholders, resources, etc.) behind the key driver (using quantification examples if possible)
4. Analyze the example of capital expenditures, recurring costs, and recurring savings
5. Provide an example of a net present value analysis

This writing utilizes a Six Sigma structure to explain the construction of an MAF using Class 1 and 2 best practices. The Six Sigma structure is an abstraction of the following Define, Measure, Analyze, Improve, and Control (DMAIC) process:

1. *Define.* Determine project objectives, scope, resources, and constraints.
2. *Measure.* Determine Critical-To-Quality (CTQ) tasks to production workflow. Obtain data to quantify process performance.
3. *Analyze.* Analyze data to identify root causes of production workflow disruptions and defects.
4. *Improve.* Intervene and change current MOM processes to improve performance through the ISA-95 transformation to a single schema across MOM systems.
5. *Control.* Implement an MAF or life-cycle management framework to maintain workflow performance through analysis of market to production workflow and system architecture.

In applying Class 1 best practices of ISA-95, "Technical Applications to Improve B2M Interface Interoperability" white papers utilize ISA-95 models to define best practice. These technical applications are the foundation for B2M and MOM interface interoperability by providing the data hierarchy and definition for interface construction. The ISA-95 MAF is then able to be constructed through Class 2 methodologies.

Step 1: Train Staff and Benchmark Technical Applications to Improve B2M Interface Interoperability (Define)

Training and benchmarks should include the following:

- B2M functional model
- B2M interface object model and attributes
- B2MML
- Part 3 MOM data flows mapped between SCOR data flows
- Proposed part 4 Level 3 MOM data flows and operations metrics
- Part 5 B2M transactions
- Proposed part 6 MOM transactions

In applying the Class 2 best practices of ISA-95, "Transformation Methodologies to Improve B2M Operations Interoperability" white papers explain the step-by-step process to improve manufacturing operations interoperability through the adoption of a single schema system architecture that supports development of an MAF for system life-cycle management for 21st-century markets.

Step 2: Structured Manufacturing Operations Assessment and Schema Migration Plan (Define, Measure, Analyze)

In order to map out the manufacturing transformation, a manufacturing operations assessment and strategy should be prepared to characterize a company's cultural and technical maturity, risks, and system road map, with a business justification for each project. The following are the minimal components of a manufacturing operations assessment:

- "As is" MOM assessment
- "To be" MOM gap analysis
- Prioritized criteria for business cases for MOM applications

Typically, an organization will not transform to ISA-95 based single schema MOM architecture in a short period of time. There is an established set of disparate terminologies, workflows, data flows, and applications that are utilized in running ongoing manufacturing operations and supply chain processes. A migration plan will involve a pre-migration step.

A good pre-migration step includes a study of operational and business drivers for a potential transformation. Since MOM covers a wide range of functions at the manufacturing level, the business drivers prioritized with regard to quickest and highest returns may assist in identifying the following information:

- MOM-related business drivers

- Priority of MOM-related business drivers

- Returns based on a net present value analysis

- Highest probability of success based on returns and current needs

The step described previously, "Prioritizes criteria for business cases for MOM applications," is required to assess the magnitude of investment and effort required in the transformation process. A sound approach includes an assessment of current MOM elements combined with a gap analysis derived from a comparison of the current MOM elements to the ISA-95 based MOM elements.

Step 3: Accelerated MOM Transformation and Application Framework Implementation (Analyze, Improve, Control)

To accelerate a manufacturer's transformation to an MAF, the following projects should be undertaken:

- Establish a project plan standard for MOM systems

- Establish standard design criteria for MOM knowledge management

- Establish schema standards for enterprise and MOM systems

- Build MOM flexible MAF

- Analyze functional segregation between enterprise and MOM applications

- Simplify functional requirement process for MOM functions and interfaces

- Simplify extended enterprise metric construction process

- Simplify event management construction process

Step 4: Life-cycle Management of MOM Application and Interfaces (Analyze, Improve, Control)

To create a sustained operations culture for continous improvement, the manufacturer should restructure it's organization to have a Manufacturing-System-Technology (MST) group to own, govern, and continuously improve the MAF. The following are tasks for this MST group:

- Use the flexible manufacturing (application) framework for change management planning and the change management process

- Simplify the workflow analytic development process

- Simplify the quality analytic development process

- Simplify skill-set mapping of super users, system and data owners, functional users, and process modelers

The biggest challenge faced by various systems and engineering departments is ensuring that the implemented MOM systems are well accepted and utilized by the user community within a manufacturing operation. An effective change management plan with process expeditors developed as part of the best practices approach alleviates some of the risks associated with nonacceptance. Educating the user community and stakeholders about the benefits of MOM applications is an ongoing, challenging cultural issue. ISA-95 MOM architectures require the collection of large amounts of data from automated as well as manual sources. Manual sources of data are typically where questionable acceptance plays a significant role in application success. User acceptance depends on a well-defined workflow combined with ongoing validation of the benefits to individual user groups and stake holders.

Manufacturing companies typically implement point solution MOM systems that are absolutely essential for short-run requirements where the politics force a bypass of some elements of ISA-95 best practice methodologies in order to ensure rapid deployment of the system. Table 1.1 highlights some differences in taking a short-term project approach with the best practices. As noted earlier, the best approach may be to follow the best practices model with elimination of a few steps in order to achieve the rapid deployment for short-term goals. Deviating from the best practice at the local project level may create a risk of nonstandardized approach but could be minimized if the deviation is kept to minimal levels.

Table 1.1. ISA-95 Business value: Holistic versus project approach		
Elements	*Generic best practices*	*Project-specific best practices*
Business drivers	Standard business and operational metrics (enterprise-wide as well as plant specific) Definition of metrics	Prioritize issues related to standard metrics and any localized metrics, based on urgency, need, returns, and so on.
Assessment standards	Function, activity, and task data flow based on ISA-95 models	As-Is MOM assessment Target versus current gap analysis
Implementation standards	Generic project plan Generic requirements process Standard design criteria for MOM ISA-95 based enterprise schema standards Common components Boundaries between different levels	Application-specific project plan Application-specific requirements Application-specific schema Common and application-specific components Interfaces based on boundaries
Life-cycle management standards	Changed management planning and process via flexible MAF Simplified workflow analytic development process Simplified quality analytic development process Skill-set mapping of super users, owners, users, and process modelers	Application-specific maintenance process Application validation of value delivered Application-specific change management process

ISA-95 MAF Enables a 21ˢᵗ-Century Lean Manufacturing Renaissance

MSA best practices will provide companies with supply chain data and software technology to apply Lean manufacturing by balancing profit, quality, and cost against each other for the On-Time Delivery (OTD) commitment decision:

- The value chain (lowest cost path to customer) to drive maximum profits
- The value stream (value-added path to meet customer's expectation) to deliver quality products and services
- The cost of product throughout its life cycle

ISA-95 describes the basis for Standard Work, which is the foundation for Lean transformation and single-piece or order flow. ISA-95 functions, tasks, and data exchanges become the Standard Work component necessary to simplify and design an MSA for a Lean supply chain. MSA that are structured for single-piece flow interact with SOA processes across the distributed supply chain of the supplier and customers to build global DDSNs. MOM solutions utilizing MSA allow appropriate organizations to make better decisions using timed, event-driven, role-based data sets that are mapped into production workflow use cases characterized by ISA-95 models and B2MML. Currently, manufacturers utilize a wide range of Lean supply chain processes that only function correctly when accurate, real-time MOM information is available and accurately directed for MTO OTD supply chains:

- *Defined customer value stream.* Benchmarking and fine-tuning production activities and quality into DDSN

- *Just-In-Time (JIT) transportation and distribution.* Coordinating "pull" logistics

- *Value-added engineering and design.* Refining product characteristics to reduce waste

- *MTO sales and marketing.* Mapping customer specification and due dates into DDSN

- *Procurement.* JIT inventory levels for replenishment and fulfillment

- *Operations.* Proactively preventing equipment breakdowns with Total Productive Management (TPM) and Overall Equipment Effectiveness (OEE) methods

To optimize the 21st-century manufacturing enterprise, companies are recognizing that production workflow, use cases (transaction sequence), and data flows must be identified, characterized in an SOA, and optimized by utilizing Lean-manufacturing and/or Six-Sigma characterization techniques. As KPI and operations metrics—with their cause-and-effect relationships (compromises) to production workflow—are developed and built into systems, ISA-95 is the enabling tool for executing this functional design efficiently. Lean MOM applications transform previously optimized Lean workflows as global markets demand change.

Lean MOM Examples

Some examples of Lean MOM include the following:

- Standard Work single schema product tracking, genealogy, and performance reporting

- Finite capacity scheduling with single-piece flow

- Theory of Constraints (TOC) for line balancing

- Utilization management using Overall Equipment Effectiveness (OEE) with resource efficiency and benchmarking to drive cultural change

- Statistical Process Control (SPC; on-line, at line, off-line) and Laboratory Information Management Systems (LIMS) for quality and workflow statistical analysis

- Role-based manufacturing portals for interdepartmental communication of real-time situations with defined event management sequences (rules)

Conclusion

This 21st-century environment requires companies to evolve their 20th-century manufacturing business model, support systems, and existing organizational practices simply to survive. They need the ability to share data and information in a secure environment so that decisions are completed more rapidly and reliably, saving time and money. The current progress of MOM applications and methods is highlighted by the following:

- Public MOM standards and methodologies are being endorsed by end users and vendors.

- MOM solutions have become part of interactive global DDSN processes.

- Vertical industry libraries of use cases and processes are being characterized using ISA-95. Resulting MOM software applications are more configurable and less a custom extension. Vendors are developing large libraries of use cases with configurable components, XML schemas, and templates toward their MSA framework for configurable interoperability.

- B2MML interfaces require much less custom interface development due to ISA-95 based libraries of configurable interfaces.

- The skill set required for MOM implementations is now recognized as a mixture of business process, IT, and manufacturing process skills.

- The return on investment for MOM solutions has been better quantified, explained, and accepted due to a large increase in repeatable MOM application sets at lower cost.

- Life-cycle costs for MOM systems are now more predictable due to an increase in tool functionality and a dramatic reduction in custom programming of interfaces and applications.

ISA-95 based MSA enables the flexible MAF of MOM systems to analyze and aggregate MOM data (e.g., capacity, capability, inventory, order and equipment scheduling) and then exchange data with Advanced Planning and Scheduling (APS), ERP, and SCM systems and the DDSNs. Developing ISA-95 best practices will provide consistency and flexibility to an extended enterprise by working interactively in real time within the supply chain to definitively determine the transformation rate to create new markets and move into them. This is 21st-century manufacturing. They enable decision making based on measurable and specific manufacturing constraints, abnormal conditions (e.g., alarms), and events. Flexible manufacturing is especially important as the United States, Europe, and the rest of the industrial world become more of a multilingual cultural melting pot. As mixing cultures, languages, foods, and fashion drive higher demand for niche and make-to-order products, manufacturers need responsive DDSNs and flexible plants to produce short, profitable production runs.

Acronym Glossary

A2F: Application To Framework
APS: Advanced Planning and Scheduling
B2M: Business To Manufacturing
B2MML: Business To Manufacturing Markup Language
CAPA: Corrective And Preventive Action
CRM: Customer Relationship Management
CTQ: Critical To Quality
DDSN: Demand-Driven Supply Networks
DFS: Design For Supply
DMAIC: Define, Measure, Analyze, Improve, Control
ERP: Enterprise Resource Planning
GAMP: Good Automation Manufacturing Practice

ISA: Instrumentation, Systems, and Automation Society
JIT: Just In Time
KPI: Key Performance Indicator
LIMS: Laboratory Information Management System
MAF: Manufacturing Application Framework
MEF: Microsoft Enterprise Framework
MES: Manufacturing Execution System
MESA: Manufacturing Enterprise Solutions Association
MOM: Manufacturing Operations Management
MRB: Material Review Board
MRO: Maintenance, Repair, and Operations
MRP: Management Resource Planning
MSA: Manufacturing Services Architecture
OEE: Overall Equipment Effectiveness
OEM: Original Equipment Manufacturers
OTD: On-Time Delivery
PLM: Product Life-cycle Management
SCE: Supply Chain Execution
SCM: Supply Chain Management
SCOR: Supply Chain Operations Reference
SOA: Service-Oriented Architectures
SOAM: SOA for Manufacturing
SPC: Statistical Process Control
TCO: Total Cost of Ownership
TOC: Theory Of Constraints
TPM: Total Productive Management
WIP: Work In Progress
XML: Extensible Markup Language

CHAPTER 2

ISA-95: A Model for Business Intelligence

Presented at the WBF European Conference, November 13–15, 2006, by

John Theron
Chief Product Officer
john.theron@incuity.com
Incuity Software Inc.
20532 El Toro Road
Suite 309
Mission Viejo, CA 92692, USA

Abstract

The ISA-95 standard, although designed to facilitate information exchange, can also be used as a model for organizing manufacturing information as part of a business intelligence solution. Adoption of the standard can then promote easier access to manufacturing information in addition to better enterprise and control system integration.

One of the challenges when implementing business intelligence solutions is to ensure that users can access the information they are looking for. Organizing the information in a model that represents equipment, material and personnel resources, products, and the production process allows users who understand the business to find the information without knowledge of the physical location or source of the data.

The elements defined in the ISA-95 standard are useful building blocks for configuring a manufacturing model. This chapter describes how users of Incuity EMI, a business intelligence solution for manufacturing, can use ISA-95 elements to build a Unified Production Model (UPM) that reflects their manufacturing environment.

The UPM organizes information from disparate sources into logical structures to help navigation and provide context. The UPM can also be directly populated from business systems and manufacturing systems using Business To Manufacturing Markup Language (B2MML) and can send data to these systems using B2MML.

Business Intelligence for Manufacturing

The investments that manufacturing companies have made in business intelligence solutions have not typically resulted in better manufacturing decisions. These business intelligence solutions have tended to embrace traditional business data sources to enhance sales, marketing, finance, and human resource decisions.

Business intelligence solutions for manufacturing require the capabilities of mainstream business intelligence solutions but also need to integrate data from control systems, historians, and Human Machine Interface (HMI) systems. With massive investments in enterprise resource planning systems, enterprise data warehouses, and enterprise historians, most organizations are reluctant to embark on yet another green field project and want to leverage the information assets contained in these systems. Federated business intelligence solutions that can leave the data in these systems, surface the data as needed, and warehouse when appropriate meet this requirement.

Incuity EMI is a business intelligence solution that helps organizations make the most of their manufacturing information. Central to Incuity is the ability to integrate information from multiple heterogeneous sources into a single unified view of the manufacturing business called the UPM.

A Unified View

The first purpose of the UPM is to integrate data from the organization's many different manufacturing data sources. These sources can include the following:

- Control systems, including Programmable Logic Controllers (PLC), Distributed Control Systems (DCS), and Supervisory Control And Data Acquisition (SCADA) systems

- Manufacturing applications, including HMI systems and historians, planning systems, maintenance management systems, Manufacturing Execution Systems (MES), and Laboratory Information Management Systems (LIMS)

- Business systems, including Enterprise Resource Planning (ERP) systems, Supply Chain Management (SCM) systems, Product Life-cycle Management (PLM) systems, and content management systems

The UPM combines and consolidates the data from these sources and presents the data via a single interface. Users do not need to know where the data actually reside. The UPM also presents a consistent interface that is independent of the data source. Users do not need to navigate the different user interfaces or understand the many access protocols for every data source.

A Manufacturing Business View

The second purpose of the UPM is to organize the data to reflect the manufacturing business. Organizing the data along manufacturing lines allows users to find information using business concepts like equipment hierarchies and material groups instead of using IT concepts like table names and tag names (Fig. 2.1).

Links and references in the UPM maintain data relationships and ensure that users can access data with context. Data from one system (e.g., a temperature measurement) become more valuable when seen in the context of data from other systems (e.g., What was being produced? Who was the customer?).

Reports, dashboards, Key Performance Indicators (KPIs), scorecards, and queries created by users across the enterprise are stored back into the UPM. This content can then be reused because the model provides a consistent interface organized along business lines. For example, a report or KPI created for a production line with one type of control system can be run against a similar line with a different type of control system by changing the parameter specifying the production line. The UPM also presents a consistent interface to applications querying the model, abstracting the complexities of individual data sources.

ISA-95 as a Model

The UPM can be developed from the ground up to represent any business operation. This does, however, represent a significant engineering effort. Manufacturing operations have much in common, and it usually makes sense to use an industry standard as a starting point.

There are many standards that define business information structures, but manufacturing standards are less common. The ISA-95 standard, developed over years by manufacturers and suppliers, has gained the most traction and is broadly applicable across different manufacturing and production environments.

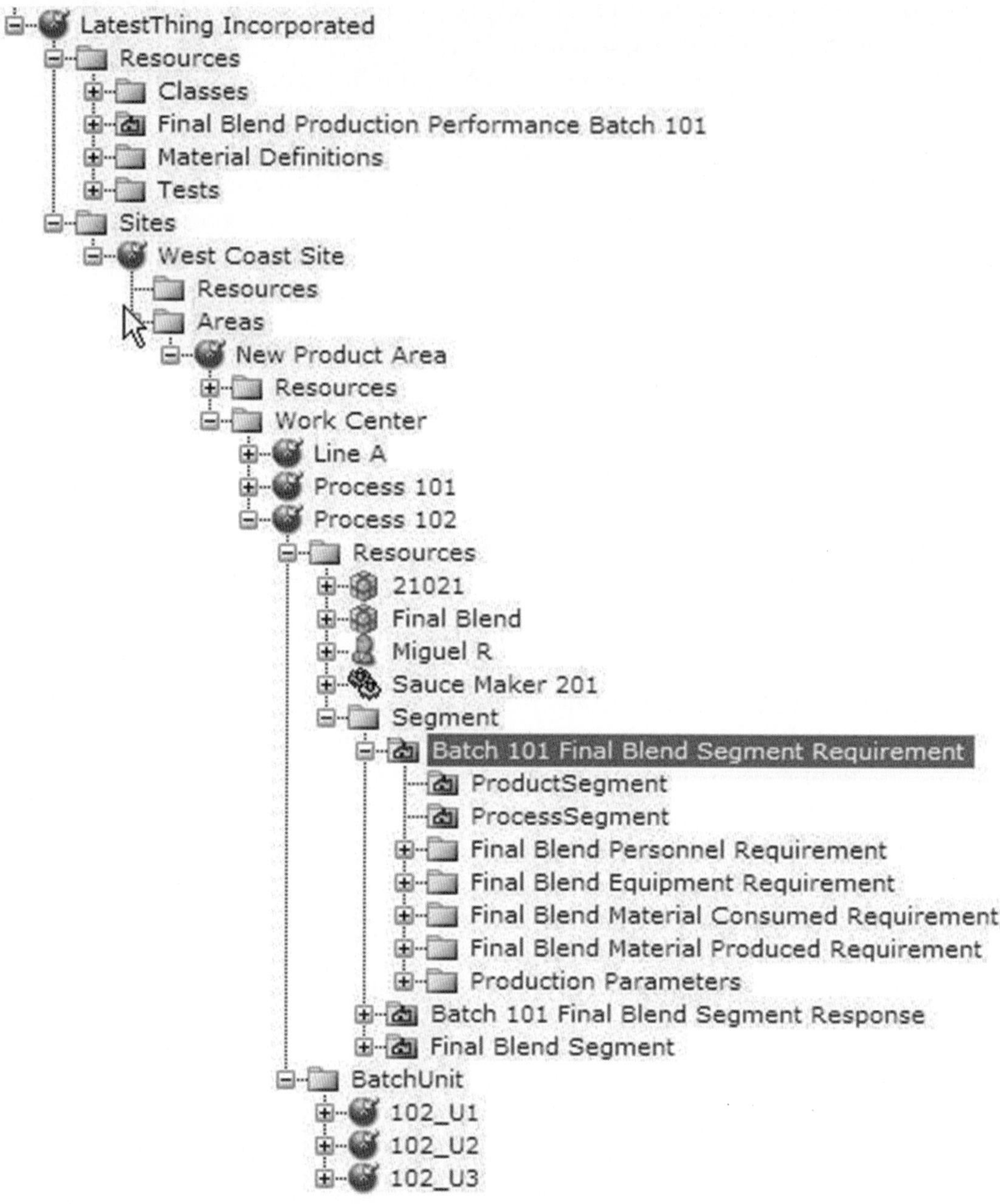

Figure 2.1. A view of the UPM.

The ISA-95 standard models all key manufacturing elements. Resource models cover the equipment, materials, and personnel needed for production. The model also includes information about the availability of these resources, how to make products, production schedules, and production performance. The ISA-95 standard therefore provides an excellent framework for organizing manufacturing information in a way that reflects the manufacturing business. The UPM includes the complete set of elements and types defined in the ISA-95 standard and B2MML.

ISA-95 Types

The UPM is based on classes called *types* and instances of types called *items*. A *pump* is an example of a type, and *Pump 101* is an example of an instance.

Types can have properties with simple data types (e.g., boolean, double, string, mime, enumerated). For example, the pump property "Rated Output" could be of type double. Types can also have reference and collection properties that define and constrain relationships with other types. The property "Pump Speed" might be a reference property pointing to a tag that measures pump speed. A connector to a control system might have a collection property containing a collection of tags. Finally, types can have operations. For example, a connector might expose an operation that synchronizes it with its data source.

The UPM contains libraries of predefined types called *packages*. Users can choose to inherit from and extend existing types or build new types from scratch to model the most complex manufacturing processes. Once a type has been defined, items (instances of this type) can be created in the namespace. The ISA-95 library introduces the following types to the model:

- Equipment types define equipment and classes of equipment. Equipment maintenance information and equipment capability tests are also defined.

- Material types define materials, material definitions, and classes of material definitions. The model includes inventory of raw, intermediate, and finished materials. Quality Assurance (QA) tests on material are also included.

- Personnel types define people, classes of personnel, and qualifications of personnel.

- Location types define a hierarchy of equipment levels and locations.

- Process segment types define groupings of the equipment, material, and personnel resources needed to carry out a production step.

- Production capability types define the capability (available, committed, or unattainable) of equipment, material, personnel, and process segment resources for a period of time.

- Product definition types define what equipment, material, and personnel resources are needed to execute a product segment (production step) for a specified quantity of product. Product definitions reference external product production rules, bill of materials, and bill of resources.

- Production schedule types specify the definition and quantity of material to be produced and the planned resources, location, start time, and end time of production (Fig. 2.2).

- Production performance types define the actual material produced and resources used and the actual start and end times of production.

At the core of the UPM are a number of types that are always present. These include tag, time period, equipment, material, personnel, and production types.

Item: [Batch 101 Final Blend Segment Requirement]

General | Properties | Operations | Security

Name	Data Type	Value
Name	String	Batch 101 Final Blend Segment Requirement
FullyQualifiedName	String	MyEnterprise.Incuity Samples.ISA95Example.La
TypeName	String	ISA95LatestThing.FinalBlendSegmentRequirem
CreatedOn	DateTime	7/10/2006 11:01:25 AM
ModifiedOn	DateTime	9/18/2006 2:10:35 PM
Description	String	Adds secret sauce to mix
IconIndex	Int32	87
Earliest Start Time	DateTime	6/14/2006 4:00:00 AM
LatestEndTime	DateTime	6/14/2006 8:00:00 PM
Duration	TimeSpan	03:20:00
Product Segment	ISA95Production.ProductSegment....	MyEnterprise.Incuity Samples.ISA95Example.La
ProcessSegment	ISA95Production.ProcessSegment....	
Required By Requested Segment Response	ISA95Production.RequiredByRequ...	Required
Final Blend Personnel Requirement	ISA95LatestThing.FinalBlendPerso...	Collection of [ISA95LatestThing.FinalBlendPers
Final Blend Equipment Requirement	ISA95LatestThing.FinalBlendEquip...	Collection of [ISA95LatestThing.FinalBlendEqui
Final Blend Material Consumed Requirement	ISA95LatestThing.FinalBlendMateri...	Collection of [ISA95LatestThing.FinalBlendMate
Final Blend Material Produced Requirement	ISA95LatestThing.MaterialProduce...	Collection of [ISA95LatestThing.MaterialProduc
Production Parameters	ISA95Production.ProductionParam....	Collection of [ISA95Production.ProductionParar

Edit...

Figure 2.2. An example of a production schedule.

ISA-95 types have been mapped to these core types so that items like tags and time periods already defined in the namespace can be properties of ISA-95 items.

Evolving the Model

The UPM supports general ISA-95 types and specific ISA-95 types. General types are free-form types that inherit from ISA-95 base types and are designed to allow users to create flexible items in the model browser. For example, users can add an item of type ISA-95Equipment.GeneralEquipment, and then add any number of properties and tags to the item. This equipment item can also contain a collection of other equipment items to reflect the equipment hierarchy. This allows users to quickly create items to model the manufacturing environment without defining new types.

The disadvantage of using the general types is that there are no constraints on the number and types of properties and tags associated with the equipment. Specific types provide these constraints and can be used to enforce model integrity. Specific types also inherit from ISA-95 base types but tightly specify their properties. For example, equipment of type pump might be defined as having only two properties—a "Rated Output" property of type double and a "Pump Speed" reference property.

Users often begin modeling with general types and progress to specific types once the model is more fully understood. Users can continue to refine and extend the model by evolving new types based on existing types and by creating new types from scratch.

Integration Services

Connectors to control systems and historians, databases, and applications collect data using a range of database connections, Web services, OPC interfaces, and proprietary interfaces. These connectors can leave the data at source and surface the data when required, cache the data, or warehouse the data depending on the data source.

Information about an equipment hierarchy might come from an asset management database. Real-time data describing the same equipment's current behavior might come from a control system. Maintenance information about the equipment may be coming from a third source—a maintenance management system. The model serves as the framework for organizing the data from the many manufacturing data sources into structures that reflect the manufacturing enterprise.

Analysis Services

Data in the model can be aggregated, summarized, and further processed on a schedule- or event-driven basis. Data movement between control systems and ERP systems can also be orchestrated. This can be done using the B2MML Web services for receiving B2MML packages from control systems and ERP systems to populate the UPM. When the B2MML packages need to be "assembled" from multiple sources, or when source systems do not have B2MML capabilities, the UPM can be populated via standard connectors. B2MML packages can be sent to control systems, ERP systems, and other consumers of B2MML.

Presentation Services

Users access the UPM to analyze the manufacturing business and create reports, dashboards, trends, and graphs via Excel or the analytic tools. These reports and dashboards are stored back into the UPM to ensure there is a single version of the truth. The reports and dashboards can be shared across the manufacturing enterprise by publishing them to a portal. Preconfigured reports against ISA-95 structures (e.g., production performance reports, schedule reports, capability reports) provide insight into the manufacturing process (Fig. 2.3).

Summary

The incorporation of ISA-95 elements into the UPM allows organizations that adopt the ISA-95 standard to benefit not only from improved ERP and control system integration, but also from a well-conceived and broadly applicable framework for organizing business intelligence for manufacturing.

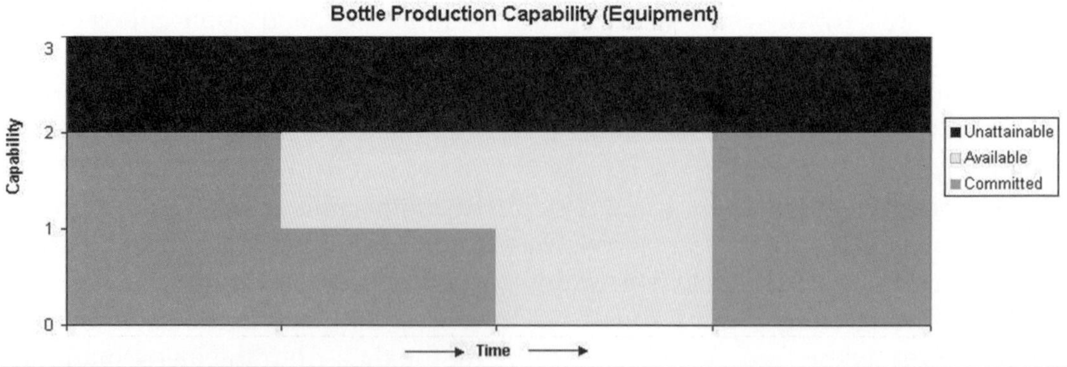

Figure 2.3. An equipment capability report.

Getting an Organization Ready for ISA-95: Designing and Implementing Plant Models within Enterprise Applications

Presented at the WBF
North American Conference,
March 24–26, 2008, by

Jack E. Greene
Associate Director
Laboratory and Process Automation Systems
jack.greene@alkermes.com
Alkermes, Inc.
88 Sidney Street
Cambridge, MA 02139, USA

Abstract

The ISA-95 approach to Business To Manufacturing (B2M) application integration defines a scalable cross-systems architecture linking applications at the business layer with applications at the control layer. These architectures can be used to streamline a wealth of business workflow models to facilitate Lean manufacturing.

Implementation of these systems is challenging because they affect all aspects of plant workflow. One of the prerequisites is to build a plant model following the ISA-88 and ISA-95 physical hierarchy models and implement a version of it within each of the applications at the business layer. This facilitates the downstream integration work and improves the utility of each individual business application.

This chapter is a case study of how a model was developed for a pharmaceutical manufacturing plant and how the model was implemented within a series of business applications. Each implementation required customization of the model to account for the strengths and limitations of each business application package, as well as the data reporting needs of the users. The specific application packages are as follows:

- Computerized Maintenance Management System (CMMS)

- Discrepancy and Corrective And Preventive Action (CAPA) tracking

- Microbiology information management system (environmental monitoring)

- Enterprise Resource Planning (ERP) and Materials Resource Planning (MRP)

Even without the Manufacturing Execution System (MES) built, the implementation of the model in each of these applications has significantly streamlined the business process by helping users to extract more meaningful information. This simplifies and reduces the work required to execute existing business processes.

Introduction to ISA-95

MESs center on the need for the transaction-based enterprise business applications (e.g., ERP, MRP, CMMS, Laboratory Information Management System [LIMS]) to transfer data among each other and to trigger changes of state based on the real-time control applications (e.g., Supervisory Control And Data Acquisition [SCADA]), Programmable Logic Controller [PLC], Distributed Control System [DCS]).

An example of the challenges that MES systems address occurs whenever a cold room temperature alarm triggers. The response to this alarm could include the following:

- The opening of a work order in the CMMS to examine and repair the cold room

- A change of state of raw material lots in that cold room from Released to Quarantined in the MRP

- The opening of an investigation within a discrepancy tracking system

Traditional models for database integration are based on a needs model. In this model, when an application needs information, an interface is built to pass the data. Over time, the business needs to integrate and pass more information among more applications, which results in a model where N applications require 2N data connectors. A portion of the connectors for a representative plant is shown graphically in Figure 3.1.

The most significant design element of the ISA-95 approach involves implementing a new application layer that sits between the business applications and the SCADA. In this scheme, the middleware transaction manager integrates with each individual business application as well as with the plant floor SCADA. This layer is used to scan applications, watch for events, and take rules-based actions in other systems based on those events.

Over time, as the business develops a greater need to integrate and pass more information among applications, the number of data connectors never exceeds 2N. This model is shown graphically in Figure 3.2. Use of this design model minimizes the number of connectors that need to be built. This has the following positive results:

- Fewer connections need to be designed, built, and maintained

- New connections and new rules associated with existing connections can be added with greatly reduced risk to existing connections and rules

- When an individual application needs an upgrade, only one other application is affected

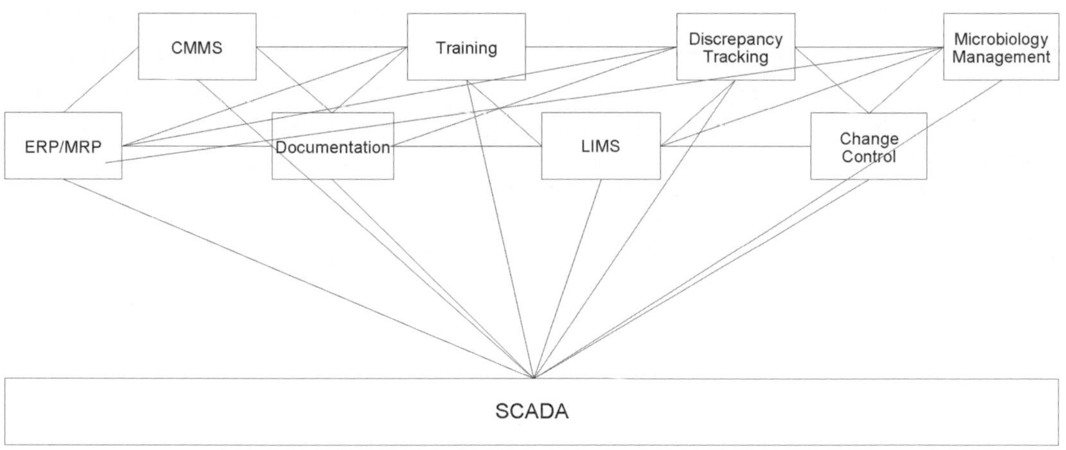

Figure 3.1. Needs model.

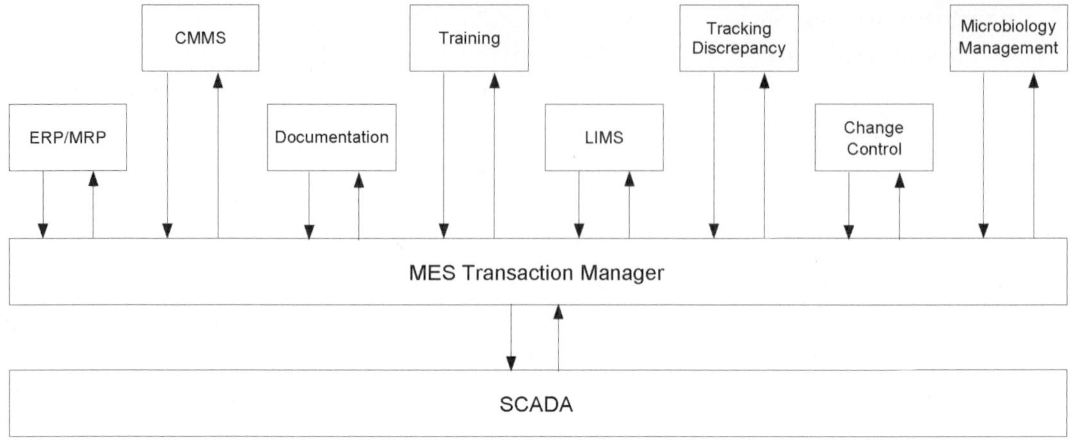

Figure 3.2. Data connectors.

- Multi-database workflow events can be designed, which increases the functionality of the system

This approach allows companies to manage a portfolio of applications, where the best application for a business need is purchased or built, knowing that the MES can be used to share the data among the portfolio in an efficient manner.

Work to Get Ready for ISA-95

While electronic workflow systems can be used to streamline and optimize processes, they only live up to their potential when the process is well understood and documented. In the case of MES, significant preparative work usually needs to be done on the existing business and SCADA applications. Without this work, the value of the MES and the types of things that it can be used for are severely limited.

For example, if a batch is going to be made on Process Cell 3 (PC3), a well-configured MES could be used to answer the following manufacturing readiness questions:

- Are all the instruments on PC3 calibrated? (query to CMMS)

- Is all the equipment on PC3 maintained? (query to CMMS)

- Is there pending calibration or maintenance work on utility systems that are called upon by PC3? (query to CMMS)

- Are there any open Deviations or CAPAs on PC3? (query to Discrepancy Tracking)

- What system design specifications and material specifications are going to be used for the batch? (query to Document Management)

- Are the users trained on the required procedures to make that batch on PC3? (query to Training)

- Are there any manufacturing restricting Change Controls on PC3? (query to Change Control)

- Are there any Microbiology Alert or Actions pending on the PC3 suite? (query to Microbiology Management)

In order to perform this check, each of the business applications needs to understand what PC3 is and its relationship to the plant's physical layout, utility layout, and controls layout. Here are four of the many critical relationships that need to be put into place:

- If the documentation system has not flagged the plant procedures that are used for manufacturing on PC3, then there will be no way to directly query the systems to find out the status of those procedures

- If the training system has not flagged who the PC3 operators are, then there will be no way to directly query to find out if they are properly trained

- If the CMMS system has not flagged which work orders for Maintenance or Calibration are for PC3, then there will be no way to directly query to find out if there are any pending

- If the CMMS system does not know what utilities PC3 needs in order to run, then there will be no way to directly query to find out if they are ready

Building the Model

First we chose to use a portion of the ISA-88.01 physical model, shown graphically in Figure 3.3. After working with the users of the systems, we discovered that in most cases we did not need to extend the model any lower than the unit level, as users rarely need groups smaller than that. We also chose to omit the enterprise level at the top, as we could assume that all tracked assets were our own or could be fit into the site level.

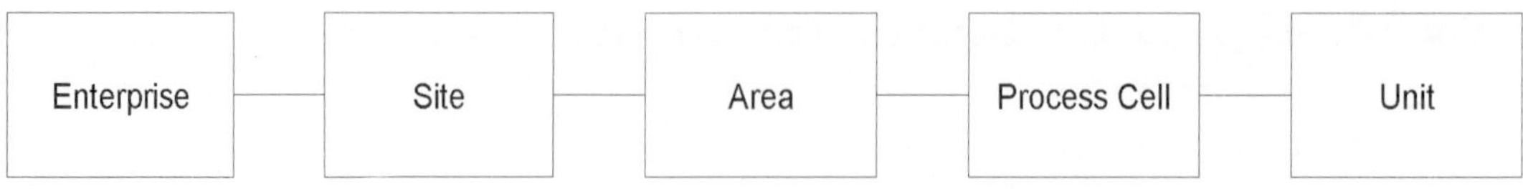

Figure 3.3. Portion of the ISA-88.01 physical model, laid on its side.

Next, we surveyed the applications to find out what level of physical model detail they could support, and then we selected the system that supported the greatest level of detail. Table 3.1 shows the applications that were used for this study. This approach will work for other brands of applications software, although the individual strengths and challenges for other brands will be different.

The Model for CMMS

We chose to build our initial model for the CMMS application, as these systems have more options for granularity. The model for all other applications would be a subset of the CMMS model. For the specific model of CMMS that was used in this study, refer to Table 3.1.

This system uses a series of hierarchal family trees to model a site. While within any given hierarchy a child may only have one parent, if a site model uses more than one hierarchy, then a child can have multiple parents. This feature is

Table 3.1. Existing applications				
	CMMS	*Microbiology data tracking*	*Discrepancy tracking*	*ERP/MRP*
Application Vendor and type	IBM Maximo	Compliance Software Solutions Corporation MIMS	Sparta Systems TrackWise	Oracle OPM
Number of model levels allowed	Unlimited	Three (software limitation)	Three (limited by our Human Machine Interface [HMI] implementation— not the software)	Three (limited by our implementation model)
Number of parents a child may have	One	One	One	One
Top level in the ISA-88 physical model	Enterprise	Site (each plant has its own instance)	Enterprise	Enterprise

critical to the way we wanted to use the application and the way we wanted the MES to query it. For example, a use point valve on the water system may be a component of the following parents:

- PC3

- United States Pharmacopeia (USP) water system

- Utility PLC

- Electrical distribution system

By having multiple hierarchies, we can implement these relationships, which will allow the downstream MES to remotely ask very complex questions. The model for the three largest hierarchies is shown in Table 3.2.

There are other hierarchies as well. The Controls hierarchy places each PLC or DCS at the Area level, each Input/Output (I/O) panel at the process cell level, and each I/O rack as a unit with children layers of cards and channels. The Power hierarchy places each main component (transformer or switchgear) at the Area level, each class of power (e.g., normal, emergency backed, and Uninterruptible Power Supply [UPS] backed) at the process cell level, and each subpanel as a unit with children layers of circuit breakers and points of use.

Table 3.2. Hierarchies			
Site hierarchies	*Area*	*Process cell*	*Unit*
Site Primary hierarchy	Bulk manufacturing	Bulk Line 1 (BL1), etc.	Standard definitions
	Filling	Fill Line 1 (FL1), etc.	Standard definitions
	QC labs	Chemistry, microbiology, stability	Chromatography systems Computerized lab instruments Pipettes Meters Balances SCADA monitored equipment
	Other labs	Each department	Each room is a unit
	Product storage areas	RT quarantined, RT released cold quarantined, cold released	Each warehouse is a unit
	Rooms with people	Each floor and building combination	Each room is a unit
	Mechanical and prep spaces	Each floor and building combination	Each room is a unit

Site hierarchies	Area	Process cell	Unit
Site Building Utility hierarchy	Heating, Ventilating, and Air Conditioning (HVAC)	Air Handling Units (AHU)	Standard definitions
	Steam	Boiler 1, 125# steam distribution, 60# steam distribution, etc.	Standard definitions
	Chilled water	Chiller 1, Cooling Tower 1, chilled water distribution, etc.	Standard definitions
	Hot water	Heat transfer skid, hot water distribution, etc.	Standard definitions
Site Process Utility hierarchy	USP water	Generation, storage, distribution	Standard definitions
	Process solvent	Loading, storage, distribution	Standard definitions
	Process waste	Collection, unloading	Standard definitions
	Aqueous waste	Collection, storage, treatment, discharge	Standard definitions
	Pure steam	Generator	Standard definitions
	Compressed air	Generation, drying, instrument distribution, clean dry air distribution	Standard definitions
	Nitrogen	Liquid, low pressure gas, high pressure gas	Standard definitions

The Model for Microbiology Data Tracking

For the specific model of microbiology data tracking that was used in this study, refer to Table 3.1. This application limits the number of layers in the model to a total of three. Because of this limitation, we opted to deploy separate database instances for each plant so that we could eliminate the enterprise and site levels in the physical model without compromising any downstream MES functionality.

The microbiology data tracking application is used to track all the sample sites in the site environmental monitoring program as well as all monitoring for all site clean utility use points. After review of the programs with the users it became clear that there were four types of room classes, as shown in Table 3.3.

Table 3.3. Microbiology room classes	
Class	*Description*
Process cell impact environmental monitoring	Findings from any room in the process cell zone affect operations on that process cell (one class per process cell)
Process cell A/B impact environmental monitoring	Findings from any room in this class affect operations on the two adjacent process cells—in this case A and B (one class per shared case)
Routine environmental monitoring	Findings from these rooms only affect operations in that room
Global environmental monitoring	Findings from any room in this class affect operations on all process cells

The microbiology data tracking model for a representative plant with two bulk process cells, two filling process cells, and a series of utility systems is shown in Table 3.4. This structure would enable the MES to be able to query microbiology data tracking for process cell–specific information. For example, to query for the status of BL1, the MES would query for Global Impact, BL1 Impact, and BL1/BL2 Shared Impact.

The Model for Discrepancy and CAPA Tracking

The next model was for the discrepancy and CAPA Tracking application. For the specific model of Discrepancy and CAPA Tracking that was used in this study, refer to Table 3.1. While this application supports almost unlimited layers, if a Web application has more than three layers of nested pulldowns, then users can lose track of what they are doing.

The Discrepancy and CAPA Tracking application is used to track all plant incidents and corrective actions for each plant in the enterprise. There had already been a field for site-level selection as well as a text field for the name of the end device (where applicable) that were required fields for all records. After reviewing the application and proposed MES functionality with the users, it became clear that while assigning each incident and CAPA to a site, area, process cell, and unit added a great deal of value, there was a diminishing return in going any deeper than that.

This discussion led to Table 3.5. This could be implemented with the application by adding a series of six buttons to the Web page that correspond to the six areas in the table. When clicked, the button would display the list of process cells in the area as a pull-down layer, with another pull-down layer for the units on that

Table 3.4. Microbiology data tracking model		
Area	*Process cell*	*Sample sites*
Bulk manufacturing	BL1 Impact environmental monitoring	BL1 dedicated suite rooms
	BL2 Impact environmental monitoring	BL2 dedicated suite rooms
	BL1/BL2 Shared Impact environmental monitoring	BL1/BL2 common suite rooms
Filling	FL1 Impact environmental monitoring	FL1 dedicated suite rooms
	FL2 Impact environmental monitoring	FL2 dedicated suite rooms
	FL1/FL2 Shared Impact environmental monitoring	FL1/FL2 common suite rooms
Routine environmental monitoring	Routine environmental monitoring	
Global Impact	Global Impact environmental monitoring	
Utilities	Air	Use points
	Nitrogen	Use points
	Clean steam	Use points
	HP water	Use points
	Hoods	Use points

process cell. This structure would enable the MES to be able to query Discrepancy and CAPA Tracking for information at any level in the model down to the unit level.

The Model for ERP/MRP

The next model was for the Oracle ERP application. For the specific model of ERP/MRP that was used in this study, refer to Table 3.1.

This application is used to track the genealogy of finished products and relate them back to the raw materials from which they were made. As materials flow along the path from the purchasing of materials to the shipment of finished goods, ERP/MRP tracks the physical location of these materials. Additionally, ERP/MRP performs financial analysis and manages the changes of state for any substance (e.g., quarantined, released, and rejected).

Table 3.5. Model for Discrepancy and CAPA Tracking		
Area	*Process cell*	*Unit*
Bulk manufacturing	Each Bulk Line	Standard definitions
Filling	Each Fill Line	Standard definitions
QC labs	Chemistry Microbiology Stability	Chromatography systems Computerized lab instruments Pipettes Meters Balances SCADA monitored equipment
Product storage areas	RT quarantined RT released Cold quarantined Cold released	Each warehouse is a unit
Site building utilities	HVAC	AHU or air distribution path
	Steam	Boiler 1, 125# steam distribution, 60# steam distribution, etc.
	Chilled water	Chiller 1, Cooling Tower 1, chilled water distribution, etc.
	Hot water	Heat transfer skid, Hot water distribution, etc.
Site process utilities	USP water	Generation, storage, distribution
	Process solvent	Loading, storage, distribution
	Process waste	Collection, unloading
	Aqueous waste	Collection, storage, treatment, discharge
	Pure steam	Generator
	Compressed air	Generation, drying, instrument distribution, clean dry air distribution
	Nitrogen	Liquid, low pressure gas, high pressure gas

After reviewing the application and proposed MES functionality with the users, it became clear that the base implementation model broke out everything by site (including the transfer of materials among sites) and handled the site material storage warehouses in a straightforward manner. As the ERP/MRP system already tracked the assignment of raw materials to batches and the timing of the start of the batch, the only piece of functionality that needed to be added was the addition of a process cell field to the system so that this assignment could be tracked.

Conclusion

The implementation of the models into the applications described in this chapter allows users to ask these systems a host of critical questions that affect the day-to-day operations of manufacturing plants:

- Were all of the instruments on PC3 calibrated?
- Was all of the equipment on PC3 maintained?
- Are there any pending calibration or maintenance work orders on utility systems that are called upon by PC3?
- Were there any open Deviations or CAPAs for PC3?
- Were there any Microbiology Alert or Actions pending for the PC3 suite?

Extending the model into other business applications such as Training, Change Control, and Document Management would allow users to be able to find out the following:

- What system design specifications and material specifications are going to be used for the batch?
- Were the users trained on the required procedures to make that batch on PC3?
- Were there any manufacturing restricting change controls on PC3?

Even without the MES in place, the ability for users to ask these types of questions enhances the value of these applications. With this structure in place, the MES implementation can run far more smoothly.

Acknowledgments

I would like to thank the following people who have helped me to plan and design the work described in this chapter:

- Don Ellwanger—Manager, Metrology Services, and Subject Matter Expert on our CMMS

- Tricia Vail—QC Microbiology Analyst II and Subject Matter Expert on Microbiology Data Tracking

- Jinna Penachio—Manager, Quality Information Services, and Subject Matter Expert on Discrepancy and CAPA Tracking

- Jennifer Strong—Manager, Materials Planning, and Subject Matter Expert on the ERP/MRP Inventory Control Modules

- Doug Brenner—Information and Technology Manager for Superior Controls and Subject Matter Expert on SCADA Integration

I would also like to thank my Supervisor Tom Harvey, Vice-President of IT, for his support, ideas, and encouragement throughout this project.

CHAPTER 4

ISA-95 Enables Flexible Discrete Manufacturing

Presented at the WBF
Make2Profit Conference,
May 24–26, 2010, by

Christian Monchinski
cmonchinski@automated-control.com
Automated Control Concepts, Inc.
3535 Route 66
Neptune, NJ 07753, USA

Abstract

The title of the ISA-95 standard discussed in this chapter is "Enterprise–Control System Integration," which implies that the standard can be implemented to allow integration of business systems to a control system layer, even where an expansive Level 3 Manufacturing Execution System (MES) implementation does not exist. A plant floor control and information system that supports each work cell independently is desirable for a more modular plant floor, as it allows for flexible, discrete manufacturing. The ISA-95 standard can be successfully applied in this environment, enabling the integration of each work cell directly into a global supply chain.

A specialty paper manufacturer has designed a modular information system to support their discrete manufacturing process. This system provides localized inventory, scheduling, and tracking for Work In Progress (WIP), operating independently for each work cell that includes work units for converting, serialized labeling, and palletizing. Lean initiatives and an increasingly global supply chain have encouraged this manufacturer to pursue a standardized interface to their existing control and information system that will remove the need for manual transcription to business systems.

This chapter will demonstrate the ability to use ISA-95 to integrate an enterprise-level system with a global view into a nimble, work cell–focused control and information system. The ISA-95 standard provided guidance during modeling, technology selection, and ultimately, implementation. The resulting system includes real-time transactions for transferring schedule, inventory status, and production performance, including finished goods. This integration has enabled the manufacturer to continue to leverage its investments and extend them to meet the demands of a globally competitive market.

Introduction

The ISA-95 standard provides both the concepts and comprehensive models for effective integration of enterprise-level systems to manufacturing operations systems. It is fairly routine today to see ISA-95 offered or implemented as an integration tool between large Enterprise Resource Planning (ERP) and equally complex MES or Manufacturing Operations Management (MOM) systems. But what if you do not have an MES? What if your MES is nontraditional, unique, or custom? Can ISA-95 be equally effective when implemented in these environments? Can an implementation of ISA-95 be "right sized" for these environments? In facilities where the Level 2 control systems are sophisticated enough to be aware of inventory, schedule, and production performance, my experience has been emphatically "yes" and for many of the same reasons that ISA-95 has found success in more traditional ERP/MES integration. By using a standard set of models and terminology, the scope and value of integrating even smaller control systems directly to ERP can be realized. The increasing availability of documentation and implementations such as Business To Manufacturing Markup Language (B2MML) support development and deployment of the standard for these implementations. As a result, the ISA-95 standard can be applied when integrating control systems directly to ERP, allowing organizations to obtain the benefits of real-time information exchange.

Background

As previously mentioned, a discrete manufacturer in the specialty paper industry embarked on a project to replicate an existing control and information system to a new green field site. I have had the pleasure to work with this innovative engineering team as a valued consultant for the past 12 years, helping them develop their Manufacturing Control System (MCS). The system had been deployed several times before and stands as a modular, self-contained work cell that includes paper converting (i.e., slitting) and robotic material handing across several work units for visual inspection, packaging, labeling, and palletizing. This converting system includes an expected assortment of Level 0, 1, and 2 components and is controlled by a Programmable Logic Controller (PLC) and Supervisory Control And Data Acquisition (SCADA). Additionally, a suite of software had been developed for local control of product configuration, detailed scheduling, machine configuration, and production and performance reporting. The Level 2 system therefore has been extended with functions usually reserved for an independent Level 3 MES, as depicted in Figure 4.1. This concept of a self-contained work cell has significant cost advantages when expanding capacity. The same equipment and automation can be quickly sourced and installed, and an experienced set of operators can be deployed to bring production on-line. Therefore, the customer's expectation is that by duplicating the existing system to this new facility, capital costs can be contained and the new facility can be brought on-line quickly.

The ISA-95 standard presents a functional hierarchy model that defines distinct operational layers within an organization. The presentation of the functional hierarchy and its boundaries are a critical first step to defining the scope of the ISA-95 standard and the content of its models and transactions. But this functional hierarchy is only one such possible model that can be encountered in any real organization. In fact, for the purposes of presenting the standard, it is by far the

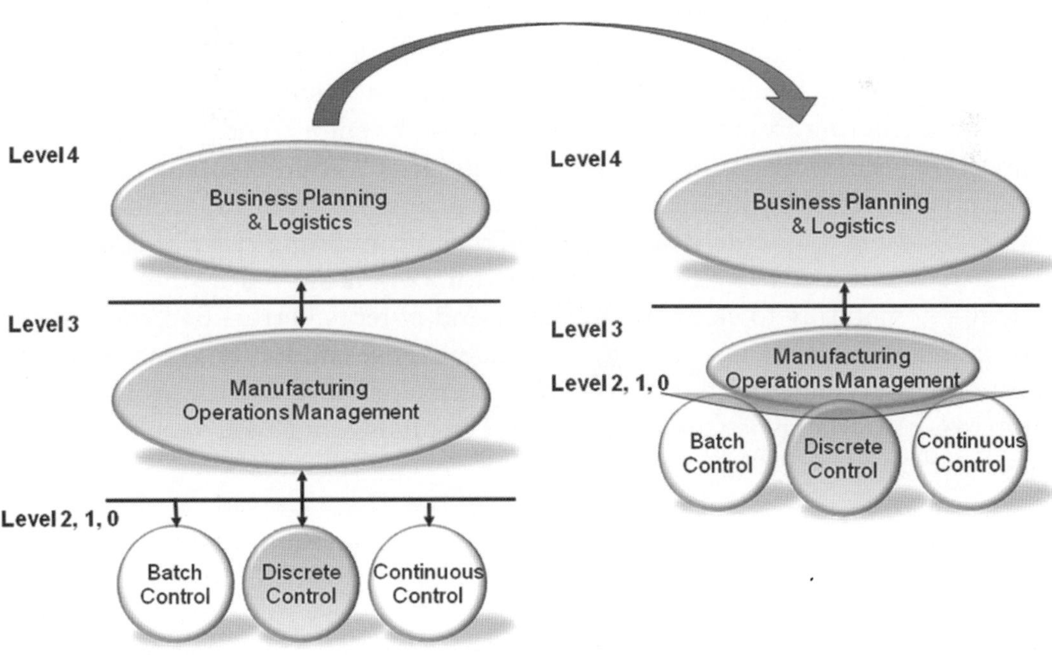

Figure 4.1. Level 2 and Level 3 overlap.

friendliest you might encounter. Real-world organizations often have business functions, systems, and people that overlap these boundaries in ways that require us to carefully consider the implementation of ISA-95. For the converting system presented here, there is effectively no Level 3 system. The compression of necessary Level 3 functions into each work cell has allowed each work cell to operate independently and identically.

A New Manufacturing Landscape

In the existing facilities where the converting system work cells exist, maintaining localized inventory, production definition, and schedule can be justified, since a common data repository supports the bulk of these duplicate manufacturing work cells. Where this system is currently in use, management has found this advantageous in that it allows them to maintain autonomy and flexibility. Inventory and schedule data are entered locally by supervisors, and production results data are transcribed into ERP using reports that are produced at the work cell. As these procedures had been the practice for more than a decade, the culture of the facility had become entrenched and little motivation existed to change the status quo.

At the new facility, however, this would not be the case. From the beginning, the new facility was envisioned as a Lean operating environment. While delivering a duplicate of the existing converting system has advantages in cost savings and implementation time, there was much legitimate concern that the converting system being delivered would require significant resources to maintain. The practice of entering in local product definitions, inventory, and schedules, all to support a single work cell, could not be justified. Of course, production definitions, inventory, and schedules do exist elsewhere—in the ERP. It did not take long to conclude that to deliver the most efficient converting system, information that resides in the ERP system would need to be transferred to the converting system automatically to eliminate the labor and potential errors that could result from manually transferring this information. Likewise and for these same reasons, information produced by the converting system for Work In Process (WIP; e.g., slitter results) and finished goods (e.g., pallet results) would need to be transferred automatically to the ERP system.

Approach to Integration

There are, of course, many ways to integrate systems. This manufacturer had prior experience using file transfers; however, these efforts were homegrown customizations between two distinct applications. As a solution, I introduced ISA-95 as an open and well-defined standard for integrating the converting system to the ERP system. There was some initial skepticism due to the "first-glance" complexity of the ISA-95 standard, but after some effort at education and a presentation of initial data mapping and technology, our engineering project team was convinced that ISA-95 was the correct approach to solving this problem. The ISA-95 standard offers two distinct advantages here:

1. *The standard is open.* Information on the standard is easily accessible. A quick search on the Internet demonstrates the standard's maturity through its ever-increasing adoption by integrators and software vendors in the marketplace.

2. *The standard is encompassing.* Unlike a point-to-point solution, the ISA-95 standard presents complete, flexible models that can be used for many different data transfers. From simple material definition exchanges to complex schedules and product definitions, the ISA-95 standard appears to handle all the possible information exchanges we will need. In this instance, the apparent "first-glance" complexity of the standard is actually an advantage.

Encountering IT

With a commitment from the engineering project team, we set out to engage the corporate information systems group and open the discussion on how we could work together to implement this interface. The IT group had its own project and resources to implement ERP at this facility in roughly the same timeline as our converting system implementation. Now, you might think that having IT engaged at this moment would be advantageous; however, the first couple of encounters with IT were anything but productive. We presented our system to IT with detailed explanations of the system's functions and needs. It was at this moment that IT declared, "We already have all of these functions and features . . . in our new release ERP." ERP owns inventory, has scheduling, and can track intermediate and finished goods—so why not put the ERP scheduling and inventory software directly on the plant floor for the operator to use?

Now I have to admit that while ERP does have inventory, scheduling, and the ability to track both intermediate and finished goods, what it lacks is the granularity and connectivity to be directly integrated to a control system. How is ERP going to download machine parameters, monitor machine values, and interpret discrete events into business data? Additionally, to dismantle the existing system and reintegrate ERP modules directly to machine-level control at this stage of the project would represent a major technical challenge. The work cell replicated here represents significant engineering investment over many years, and fortunately my customer recognized that.

Our engineering team formed a separate vision for integration. In effect, the converting system was to be treated as if it were in a bubble. The rule applied was simple: the bubble is an interface that can only be penetrated using standard interfaces. The ISA-95 standard, presented as the tool for integration between the ERP system and converting control system, fit nicely into this vision. With IT resources at a premium, I was able to use ISA-95 to organize requirements and present initial data mappings between the converting system and the ERP system. This preliminary work gave a clear picture of the effort needed to make this interface a reality and gave the engineering team a head start in the integration effort. Impressed with the work done to this point, IT directed that this standards-based integration approach would be the best approach going forward and committed their resources to the effort.

Mapping Exercise

In order to begin the implementation of an ISA-95 integration effort, a detailed mapping exercise was conducted. The WBF's recently published B2MML Mapping Sheets were used as guidance for this mapping exercise. The mapping sheets, a collection of Excel spreadsheets, provide a visually intuitive way to perform a side-by-side comparison of data required by one system and consumed by another, using ISA-95, specifically B2MML, for transference. The exercise was executed by first presenting the converting system's information capabilities and then mapping each requirement through the ISA-95 model to the source or destination data within the ERP system.

Necessary information exchanges were identified between the ERP system and the converting system, including a production schedule, intermediate material inventory for inbound paper rolls to the converting system, a production response that includes intermediate WIP, a final production response that represents the creation of new finished goods inventory, and product definitions to define which products could be made with the converting system. In total, seven information exchanges were identified, as depicted in Figure 4.2.

A production schedule is required, as the converting system maintains a local production schedule. The converting system's production schedule is more akin to production dispatch and execution, in that it is singularly focused on the converting

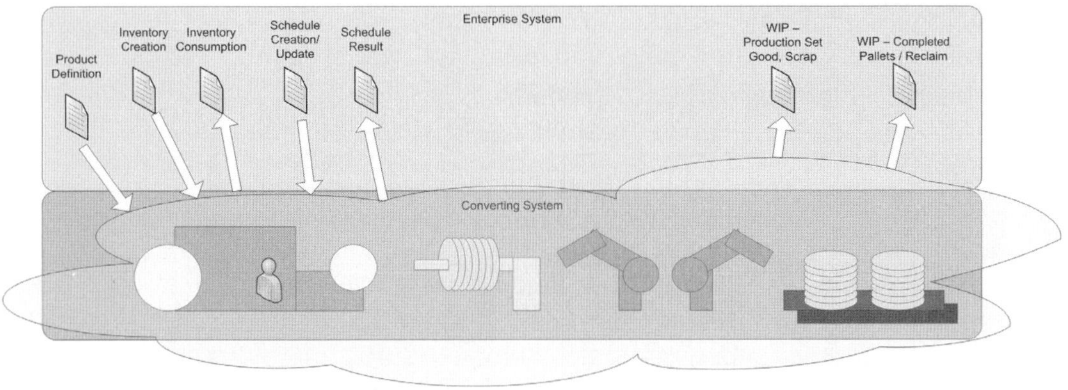

Figure 4.2. Required information exchanges.

system; however, it does have the ability to queue production requests. The critical piece of the production schedule in the converting system is that by selecting and activating a schedule, the product parameters from a stored product configuration are applied to the machine's real-time configuration. A periodic transfer of the current production schedule is sent by the ERP system to the converting system, where the converting system will process the new production schedule and store it locally. This transferred schedule includes production requests for a given period of time, as determined by the ERP system. An example of the data mapping for the production schedule is shown in Table 4.1.

Within the converting work cell, the inventory is unprocessed paper rolls. In other plant locations, where the paper is created in the same mill that it is converted, these paper rolls are handled as an intermediate product stage. It follows, therefore, that the converting system can effectively maintain a localized inventory of paper rolls, including key quality attributes that are necessary for determining the suitable end products that can be made from the base paper. This new facility will implement converting with sources of paper from other domestic and international corporate sites, as well as outside manufacturers. This new, complex supply chain requires that inventory now be maintained and owned by the enterprise system. The converting system still requires a local snapshot of inventory that represents the available inventory of paper rolls. Again, a periodic transfer of current paper roll inventory is published by the ERP system to the converting system. This publication can include a complete inventory snapshot or a subset of inventory (e.g., new inventory), as determined by the ERP system.

In the analysis of paper roll inventory represented as incoming material, one key issue that surfaces surrounds representing material attributes. The converting system expects a paper roll to be dissected into multiple grades. This

Table 4.1. Example production schedule mapping			
Converting system database		*ISA-95 element*	*Mapping/transform action*
Field	*Meaning*		
Table [SCHEDULE]			
[WO_NUMBER]	Work order number (from Oracle R12)	`yncProductionSchedule/DataArea/ ProductionSchedule/ProductionRequest/ID`	From ERP
[FURNISH_CODE]	Product item code (finished goods)	`/SyncProductionSchedule/DataArea/ ProductionSchedule/ProductionRequest/ SegmentRequirement/MaterialRequirement/ MaterialLotID`	From ERP
[ORDER_DUE_DT]	Date and time of expected order completion	`/SyncProductionSchedule/DataArea/ ProductionSchedule/ProductionReq   me`	From ERP
[AMOUNT_RQD]	Target production required	`/SyncProductionSchedule/DataArea/ ProductionSchedule/ProductionRequest/ SegmentRequirement/MaterialRequirement/ Quantity`	From ERP
[SCHEDULE_ID]	Internal database key		Assigned at time of creation, not part of the transfer

"classification" of inventory is a quality function that occurs after paper receipt, as depicted in Figure 4.3, and is necessary in order for the converting system to check that the appropriate material is being used to make product.

In order to ensure that all material attributes are contained in the information transfer from the ERP system, quality information about inventory will need to be transferred into the ERP system first. Then the material information can be transferred to the converting system, but only after the material is in a status of "released from quality control." This requires an additional integration effort between the quality system and ERP system, adding additional complexity and time to the overall effort.

Two production responses that can come from the converting system are (1) information about each intermediate set of paper rolls that is produced by the converting system and (2) the creation of a pallet. The amount of data collected from the converting system far exceeds the amount that the ERP system requires. What is required by the ERP system is a birth-of-pallet transaction that includes a unique pallet number (i.e., license plate) and the number of items on the pallet. Additionally, the converting system provides information to the ERP system that uniquely identifies the origin of each converted paper roll on the pallet. The converting system also supplies totalized information on good converted paper rolls and those destined for reclaim or a landfill, as a WIP Scrap/ Reclaim transaction.

Finally, the converting system requires a definition of each product. This is by far the most complex transaction to map, as the data stored in the ERP system do not clearly match the data required by the converting system. Each product definition required by the converting system includes product specifications, tolerances, customer information, labeling requirements, packaging requirements, and operator instructions. Table 4.2 shows a summary of each of the information transfers and the identified ISA-95 objects used.

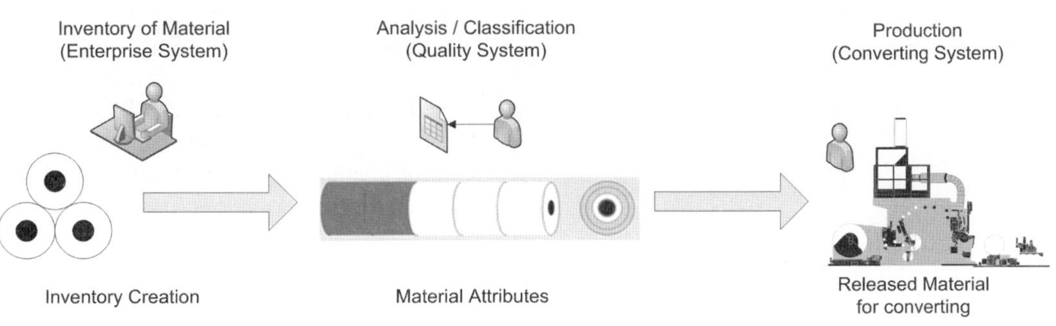

Figure 4.3. Applying quality attributes to inventory.

Table 4.2. Data transfers implemented using ISA-95 and B2MML		
Information transfer	*Implemented using*	*Implementation details*
Product Definition Transfer	`SyncProductDefinition B2MML-V0401- ProductDefinition.xsd`	This is a message sent from the ERP system to the converting system. Separate product segments have been used to define different aspects of the product definition, including the paper grade requirements, specifications for each individual paper roll, labeling requirements, and palletizing requirements. It is by far the largest of the data transfers. Upon receipt, the data are transformed and updated into seven separate relational database tables in the converting system.
Material Transfer (Paper Roll Available Inventory)	`SyncMaterialLot B2MML-V0401-Material. xsd`	MaterialClass and MaterialDefinition objects are not exchanged, as the material transfer required to the converting system is always for "paper rolls." However, different material definitions could have been implemented for different classifications of paper, and the interface may be expanded in the future to achieve this end.
Material Transfer (Paper Roll Consumption)	`SyncMaterialLot B2MML-V0401-Material. xsd`	This response is sent from the converting system back to the ERP system that relates the current paper roll diameter, calculated weight, and the local "consumed" status of the roll. One or more paper rolls may be defined as individual material lots within the transfer.
Production Schedule Transfer	`SyncProductionSchedule B2MML-V0401- ProductionSchedule. xsd`	This production schedule is sent periodically from the ERP system. Within the production schedule, multiple production requests can be contained for each work order. All are for a single segment requirement— paper converting.

Table 4.2. Data transfers implemented using ISA-95 and B2MML *(continued)*		
Information transfer	*Implemented using*	*Implementation details*
Production Performance Transfer (Schedule Item Completed)	`SyncProductionPerformance B2MML-V0401- ProductionPerformance .xsd`	This response is sent from the converting system back to the ERP system. It contains a production response that ties back to the production request ID provided in the production schedule transfer. The response type is used to indicate the production request status.
Production Performance Transfer (Intermediate Production)	`SyncProductionPerformance B2MML-V0401- ProductionPerformance .xsd`	This response is sent from the converting system back to the ERP system. In this response we use material actual with a predefined material class ID and material definition ID that identify the material actual as an intermediate paper "set." Three material actual properties are contained within the material actual that define the count of scrapped, reclaimed, and sampled material removed from the paper set.
Production Performance Transfer (Finished Goods Production)	`SyncProductionPerformance B2MML-V0401- ProductionPerformance .xsd`	This response is sent from the converting system back to the ERP system. Each pallet of material ejected from the converting system represents finished goods. In this response we again use material actual with a predefined material class ID and material definition ID that identify the material actual as WIP completed. The transition to finished goods actually happens in the ERP system; however the material lot ID of each material actual is the pallet license plate number, used by the ERP system for unique identification of the pallet. Material actual properties are used to identify each paper set on the pallet.

Implementing ISA-95

No discussion of an integration project would be complete without a survey of the implementing technology. Of course the ISA-95 standard is an integral part of the solution, but the ISA-95 standard needs to be "implemented," and there are many ways of accomplishing this.

The data-mapping exercise allowed us to identify the relevant ISA-95 object and to represent the data that will be exchanged. Following the initial data mapping exercise, each data exchange was reviewed for content and a determination was made for the frequency and transaction type required. For each of the data exchanges, it was determined that a publish-subscribe model would be adequate. The publish-subscribe model, presented in ISA-95 part 5 and shown in Figure 4.4, is a loosely coupled integration that allows the system's exchanging information to post (i.e., publish) information asynchronously to one or more recipients (i.e., subscribers).

The information transfers between the ERP system and the converting work cell are bidirectional. Despite the real-time operation of the converting work cell, these information transfers do not (necessarily) need to be real time and are not implemented as such in this case. The frequency of refresh is dependent on the need for the data. For example, the plant has the option to update a production schedule multiple times during the day, whereas product definitions, once configured, change less frequently.

Figure 4.5 shows the information transactions between the various systems in a simplified view. The view is simplified in the sense that the data represented in the various ERP system data stores must be aggregated and then transformed to B2MML syntax (a detail not shown here). The converting system side of the figure is a nearly complete detail of the resulting transfer from the converting system's perspective. Converted work-cell data reside in a series of tables in a local, relational database.

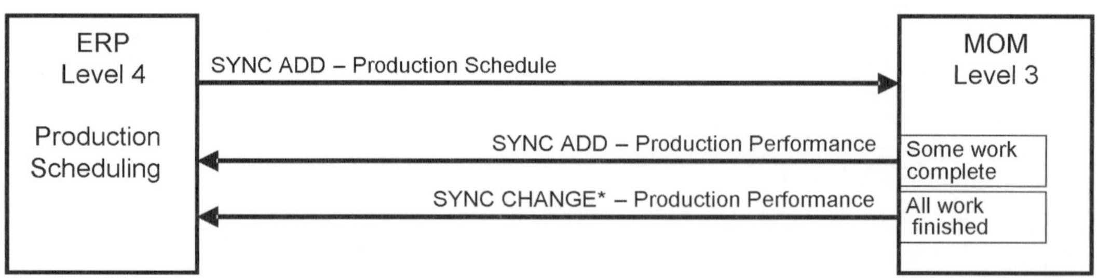

Figure 4.4. ISA-95 part 5 publish-subscribe model for production schedule and production performance.

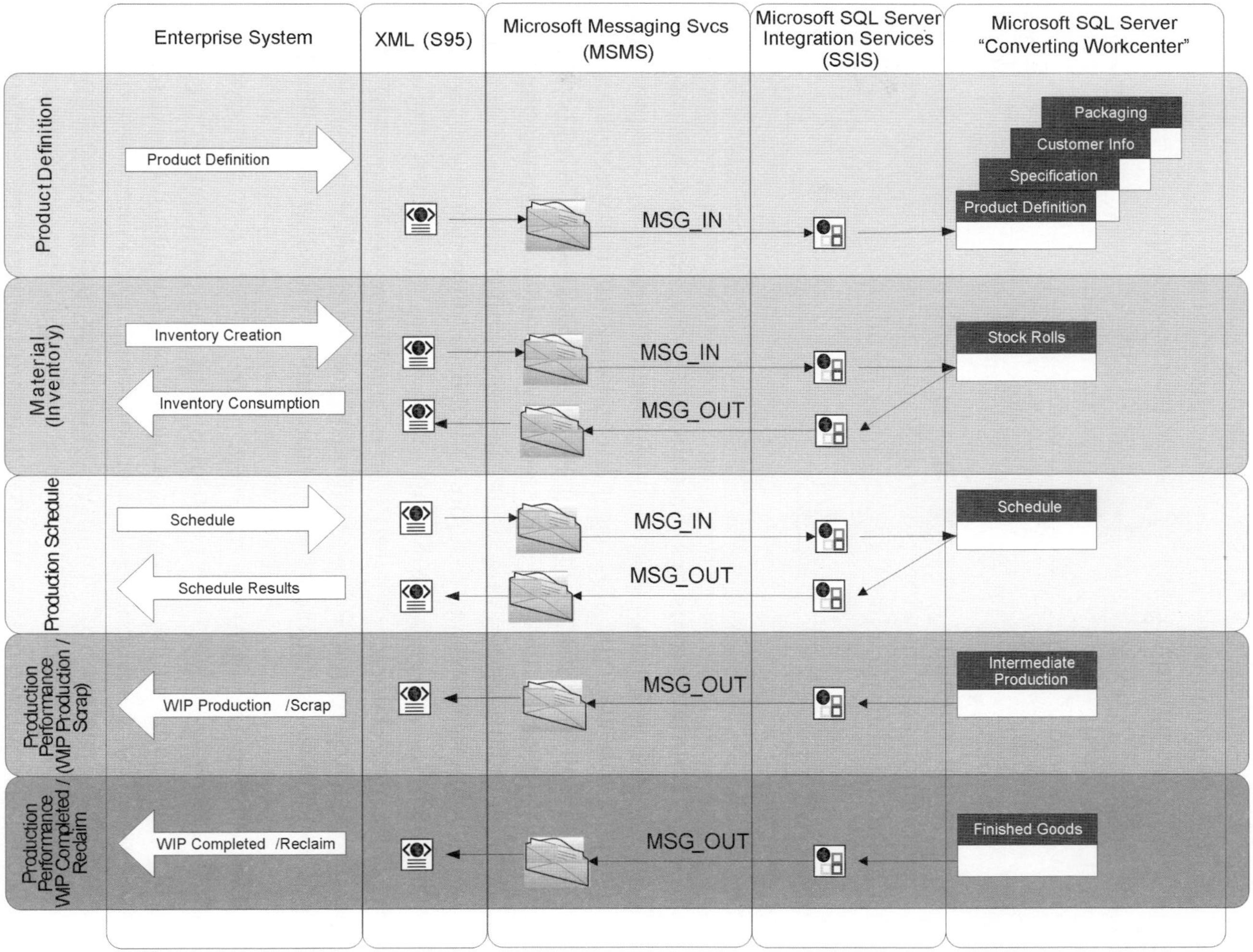

Figure 4.5. Information exchange.

In order to facilitate a publish-subscribe model for information transfer, Microsoft Messaging Services (MSMS) can be implemented to provide mailbox locations for sending and receiving information. Within each of the mailboxes are Extensible Markup Language (XML) documents representing the transferred information. The B2MML implementation involves the creation of Extensible Stylesheet Language Transformations (XSLT) to generate the B2MML documents from data in the converting system, as depicted in Figure 4.6. A similar effort to develop XSLT documents for the ERP system is also undertaken.

From the converting system, receiving a B2MML document containing production definition, production schedule, or material information (e.g., inventory) involves removing the incoming message from the inbound mailbox, transforming the B2MML using XSLT, and placing the information into the appropriate target database tables. This involves the development of interfacing logic to the mailbox, application of the stylesheet transformation, and finally reconciliation of the logic to the local database tables. While more sophisticated systems may accomplish this with Web services, Service-Oriented Architectures (SOA), and the like, the system here was adequately served using Microsoft SQL Server Integration Services (SSIS) and a series of .NET Common Language Runtime (CLR) extensions to the database server. An example of this logic is shown in Figure 4.7.

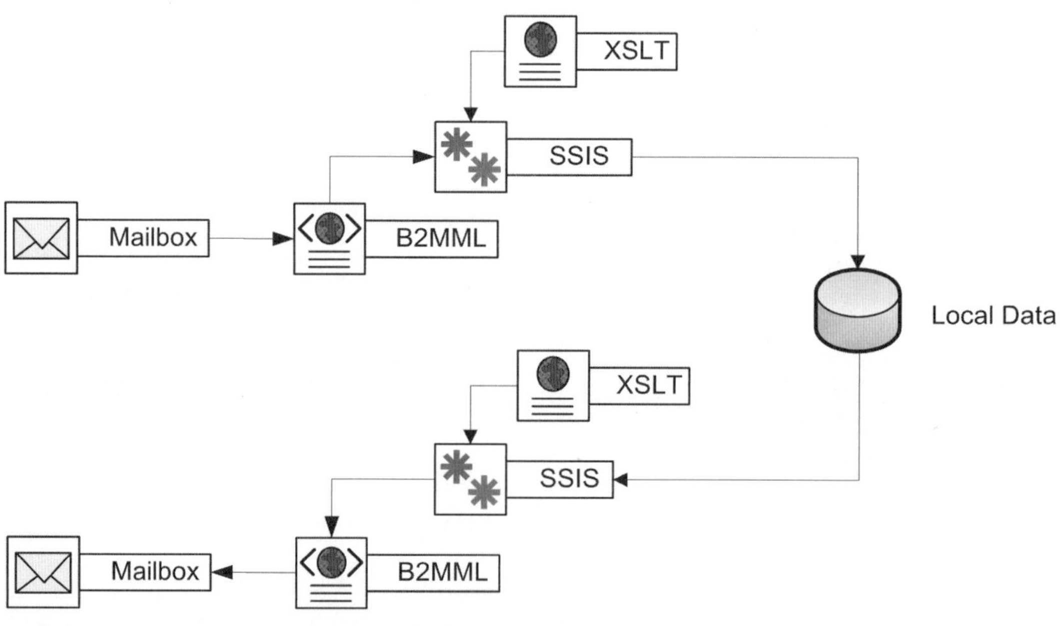

Figure 4.6. Message publish and subscribe using B2MML.

```
// Simple receive of message, string type assumed
Public ReceiveQString(String sQueueName)
{
    MessageQueue msgQ = new MessageQueue(sQueueName);
    String msgText;
    Message myMessage = msgQ.Receive();
    Type targetTypes(0) = GetType(String);
    myMessage.Formatter = new XmlMessageFormatter(targetTypes);
    msgText = String (myMessage.Body);
}
// Transform xml document. Imports xslt and produce result xml document
Public TransformXSLT(ByVal sourceDoc As String,
ByVal resultDoc As String,
ByVal xsltDoc As String)
{
    XPathDocument myXPathDocument = new XPathDocument(sourceDoc);
    XslTransform myXslTransform = new XslTransform();
    XmlTextWriter writer = new XmlTextWriter(resultDoc, null);
    myXslTransform.Load(xsltDoc);
    myXslTransform.Transform(myXPathDocument, null, writer);
    StreamReader stream = new StreamReader (resultDoc);
}
```

Figure 4.7. Example receipt of XML message from a message queue and transformation with XSLT.

Because the target format of the converting control system is a set of relational database tables, the XSLT transformation is a straightforward mapping of the necessary B2MML fields to XML-persisted data sets that represent the relational database tables of the converting system. Figure 4.8 shows an example of a B2MML for SyncProductionPerformance to report a schedule item completion. Figure 4.9 depicts source data for this schedule item from the converting system database mapped to the SyncProductionPerformance B2MML using an XSLT generation tool. Figure 4.10 shows an example of the resulting XLST that would be used to create the B2MML from the source data.

```xml
<?xml version="1.0" encoding="utf-8"?>
<SyncProductionPerformance
    xmlns="http://www.wbf.org/xml/B2MML-V0401"
    xmlns:xsi="http://www.w3.org/2001/XMLSchema-instance"
    xsi:schemaLocation="http://www.wbf.org/xml/B2MML-V0401
B2MML-V0401-ProductionPerformance.xsd">
    <ApplicationData>
        <Sender>
            <LogicalID>123431</LogicalID>
        </Sender>
        <CreationDateTime>2010-01-01 06:30:59</CreationDateTime>
        <Signature></Signature>
    </ApplicationData>
    <DataArea>
        <Sync>
            <ActionCriteria></ActionCriteria>
        </Sync>
        <ProductionPerformance>
            <ID>20100301-3</ID>
            <Description>SWM Daily Production</Description>
            <Location>6</Location>
            <ProductionScheduleID>2010-03-01</ProductionScheduleID>
            <StartTime>2010-03-01 00:00</StartTime>
            <EndTime>2010-03-01 23:59</EndTime>
            <EquipmentElementLevel>0</EquipmentElementLevel>
            <ProductionResponse>
                <ID>26416 - 2.1</ID>
            <ProductionRequestID>26416-2.1</ProductionRequestID>
                <Description>XYZ Australia</Description>
<ProductProductionRuleID>F10615EE272D0</ProductProductionRuleID>
                <Location>6</Location>
                <StartTime>2010-03-01 11:00</StartTime>
                <EndTime>2010-03-01 15:00</EndTime>
                <ResponseType>COMPLETE</ResponseType>
            </ProductionResponse>

        </ProductionPerformance>

    </DataArea>
</SyncProductionPerformance>
```

Figure 4.8. Example of an XML file for production performance sync for a schedule that is complete.

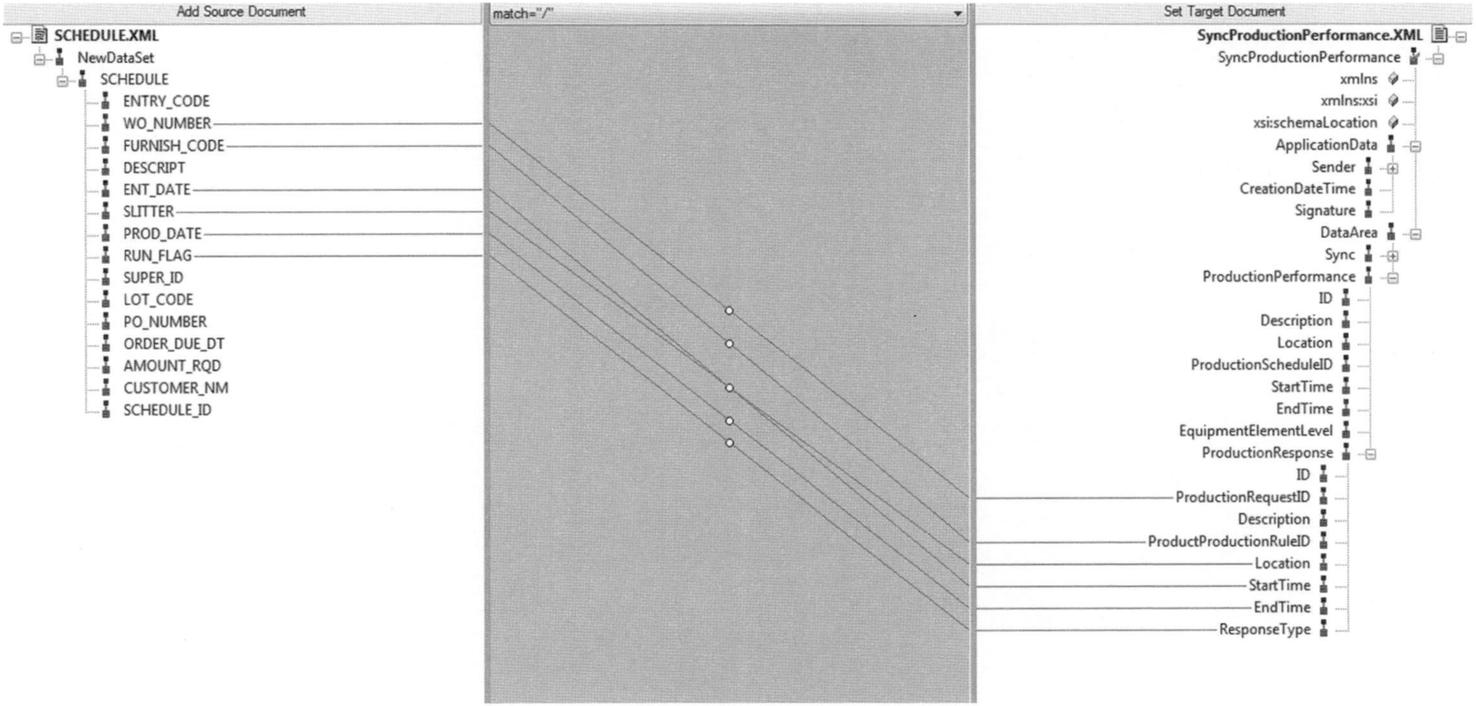

Figure 4.9. Graphical mapping of converting system data to B2MML, production performance.

```xml
<?xml version='1.0' ?>
<xsl:stylesheet version="1.0" xmlns:xsl="http://www
.w3.org/1999/XSL/Transform">
    <xsl:template match="/">
        <SyncProductionPerformance xmlns="http://www.wbf
.org/xml/B2MML-V0401" xmlns:xsi="http://www.w3.org/2001/
XMLSchema-instance">
            <DataArea>
                <ProductionPerformance>
                    <ProductionResponse>
                        <ProductionRequestID>
                            <xsl:value-of
select="NewDataSet/SCHEDULE/WO_NUMBER"/>
                        </ProductionRequestID>
                        <ProductProductionRuleID>
                            <xsl:value-of
select="NewDataSet/SCHEDULE/FURNISH_CODE"/>
                        </ProductProductionRuleID>
                        <Location>
                            <xsl:value-of
select="NewDataSet/SCHEDULE/SLITTER"/>
                        </Location>
                        <StartTime>
                            <xsl:value-of
select="NewDataSet/SCHEDULE/ENT_DATE"/>
                        </StartTime>
                        <EndTime>
<xsl:value-of select="NewDataSet/SCHEDULE/PROD_DATE"/>
                        </EndTime>
                        <ResponseType>
                            <xsl:value-of
select="NewDataSet/SCHEDULE/RUN_FLAG"/>
                        </ResponseType>
                    </ProductionResponse>
                </ProductionPerformance>
            </DataArea>
        </SyncProductionPerformance>
    </xsl:template>
</xsl:stylesheet>
```

Figure 4.10. Example resulting XSLT for production performance.

Conclusion

Implementing ISA-95 to integrate ERP system data directly to Level 2 systems is achievable. The ISA-95 standard delivers an impact immediately by allowing information systems personnel and engineering personnel to communicate requirements, scope, and technical detail quickly and accurately. The business value is apparent through results such as reduced manual data transcription and real-time process response, and ISA-95 is used to deliver this value with an open integration platform. The ISA-95 standard, as implemented in the WBF B2MML schemas, can be used effectively to integrate a Level 2 system to the enterprise, using little more than the standard technology set already familiar to automation engineers. The result is a preservation of engineering investment in the functionality of this autonomous converting system and an enhanced manufaturing operation brought about by integration to an ERP system with global knowledge of inventory and customer demand.

CHAPTER 5

Mapping ISA-95 Production Schedules, ISA-95 Production Performance, and ISA-88

Presented at the WBF North American Conference, March 24–26, 2008, by

Hernán Felipe Bolaños Cruz
Industrial Automation Engineer
hbolanos@unicauca.edu.co
University of Cauca
Street 5 DE No. 9B-26
Popayán (Cauca), 057, Colombia

Juan Manuel Velásquez Vélez
Industrial Automation Engineer
jvelasquez @unicauca.edu.co
University of Cauca
Street 26 CN No. 6B-35
Popayán (Cauca), 057, Colombia

Oscar Amaury Rojas Alvarado
Electronic and Telecommunications Engineer
orojas@unicauca.edu.co
University of Cauca
Street 5 No. 4-50
Popayán (Cauca), 057, Colombia

Abstract

Automation concepts have spread to all levels of organizations within the industrial sector, expanding the concepts and requirements beyond the control systems' and the machines' reach, to the point of affecting the business procedures of the company. Therefore, to get an integral automation production process, it is necessary to keep in mind various aspects such as the process itself, the company's organizational structure, the equipment, and the management of the supplies. The specification and the definition of such aspects are even more complex, due to economic policies and globalization. There is an imperative to get the most out of the information flow of the company's business processes in order to maximize productivity and competitiveness.

In this chapter, the study of the information exchange between the batch management software using Batch Markup Language (BatchML) and the Manufacturing Execution System (MES) using Business To Manufacturing Markup Language (B2MML) is presented, having as a base the common information mapping between the manufacturing execution levels and the control level.

This chapter will identify the BatchML fields that are related to the information represented by the B2MML, according to the Production Performance model and the Production Schedule model. The relationship between the Production Schedule model and the batch management information is presented, as is the information delivered by the batch management software to the MES level in the Production Performance model.

The information flow for the Production Schedule and Production Performance documents, which are exchanged between the manufacturing and the control level, is defined using the structure proposed by the ISA-88 and ISA-95 standards. This is part of the information exchange problem between batch management systems and MES systems. Furthermore, as the information exchange definition in this chapter progresses, it becomes certain that a generic standard interface can be obtained through dynamic document exchange using B2MML and BatchML formats.

Introduction

Automation concepts have spread to all levels of organizations within the industrial sector. These concepts have spread beyond the reach of the control systems and the machines, affecting the business procedures of the customers and their suppliers in the areas of product quality management, systems maintenance, data instrumentation and acquisition, statistical production control, resource and supply planning, purchasing, and so on.

Therefore, in the industrial automation processes, it is necessary to obtain complete integration of information, starting with lower process levels, going up through the control level, and eventually reaching the management level. In most companies, the business management and production control systems are not separated by a lack of technology, but by inappropriate management of the information.

In this task—which at first was thought to be easily carried out—numerous difficulties have been found in practice, not only in the technological and economical field, but also in the technical and conceptual fields. For this reason, it is necessary to apply international standards to the development, with the objective of obtaining a simpler integration between the business management level and the production control level.

Mapping ISA-95 to ISA-88 for Production Schedule and Performance

The development of the ISA-88 and ISA-95 standards has been most relevant to clarifying communication between business and production systems. ISA-88 defines a set of terms and models used to define the control requirements for batch manufacturing processes. The standard describes how to guarantee detailed information exchanges for transmission of product recipes and production and control activities. It also defines data models for an efficient interface between the involved subsystems in batch control. ISA-95 defines a common language for integration between the production execution and business management systems. To achieve this objective, there is a superposition of models and terminology for the lower level in ISA-95 and the upper level of ISA-88. Therefore the components in the MES level must handle both the B2MML and BatchML languages to get an efficient and controlled information flow between the management level and the control level, respectively.

In this scenario the middleware applications acquire great importance, since they guarantee a standard and generic interface throughout the document exchange. As shown in Figure 5.1, the necessary middleware items in the ISA-95 standard are the Enterprise Resource Planning (ERP) middleware and middleware for the communication of the MES system with the batch management software applications. The ERP middleware transforms the information from the ERP's proprietary format to a B2MML structure and vice versa. The middleware for the batch system transforms the information delivered by the batch management software in a BatchML structure to a B2MML structure and vice versa.

The BatchML documents specify the data types and elements commonly found in batch control applications that are used for information exchange for the companies, systems, or specific document applications that can be derived from the BatchML documents. The B2MML documents, as defined by the WBF, are an Extensible Markup Language (XML) implementation from the ISA-95 standard

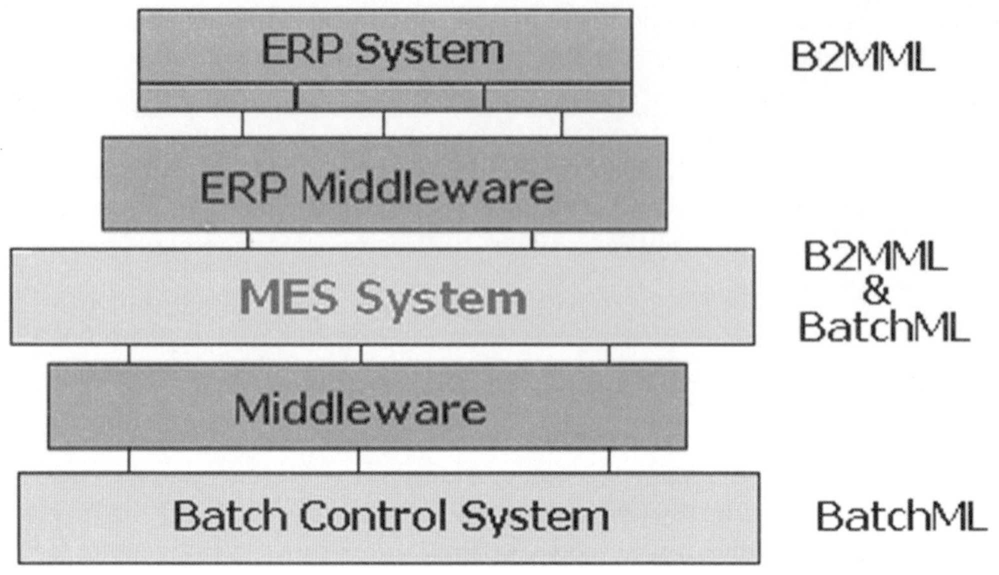

Figure 5.1. ISA-95 and ISA-88 integration scenario.

family and are designed to be a common data format for the link between ERP and MES.

After carrying out the general structure analysis of the BatchML documents and the B2MML documents corresponding to the Production Schedule and the Production Performance, a mapping of the information handled by the batch management in its BatchML documents and the MES in its B2MML documents (and the common information related to these documents) can be defined. First, the Production Schedule document fields in B2MML that have correspondence with the BatchML document elements are identified. Subsequently, the BatchML document fields that have correspondence with the Production Performance elements according to the B2MML structure are identified.

Analysis

As a result of this analysis, a portion of the information mapping between the BatchML documents and the Production Schedule and Production Performance documents to B2MML is shown (Table 5.1).

The first column of Table 5.1 shows all the fields that belong to the BatchML that are related to the analyzed documents in B2MML. The second column shows the information that represents the relationship between the information flow from BatchML to B2MML, according to the Production Performance model. The

Table 5.1. Mapping of the BatchML document to Production Schedule and Production Performance in B2MML		
BatchML	*Production Performance (BatchML to B2MML)*	*Production Schedule (B2MML to BatchML)*
MasterRecipeType		
IDType	ProductionSchedule/ID ProductionPerformance/ ProductionSchedule/ID	ID ProductionSchedule/ID
ProductNameType: Header/ ProductName	DescriptionType: MaterialProducedActual/ Description	DescriptionType: MaterialProducedRequirement/ Description
NominalType: Header/ BatchSize/Nominal	QuantityStringType: MaterialProducedActual/ Quantity	QuantityStringType: MaterialProducedRequirement/ Quantity
ParameterType: Formula/ Parameter/ID	ParameterProductionID: SegmentResponse/ ProductionData/ID	ParameterType: ProductionParameter/ Parameter/ID
DescriptionType: Formula/ Parameter/Description		DescriptionType: ProductionParameter/ Parameter/Description
ValueStringType: Parameter/ Value/ValueString	ValueStringType: ProductionData/Value/ ValueString	ValueStringType: ProductionParameter/ Parameter/Value . . .
DataType: Parameter/Value/ DataType	DataType: ProductionData/ Value/ValueString	DataType: ProductionParameter/ Parameter/Value . . .
UnitOfMeasureType: Parameter/Value/ UnitOfMeasure	UnitOfMeasureType: ProductionData/Value/ UnitOfMeasure	UnitOfMeasureType: ProductionParameter/ Parameter/Value . . .
EquipmentElement		
IDType:	EquipmentIDType: Location/EquipmentID	EquipmentIDType:Location/ EquipmentID
EquipmentElementLevelType:	EquipmentElementLevelType: Location/ EquipmentElementLevel	EquipmentElementLevelType: Location/ EquipmentElementLevel
BatchListEntry		
IDType:	IDType: SegmentResponse/ ID	IDType: SegmentRequirement/ ProductSegmentID

Table 5.1. Mapping of the BatchML document to Production Schedule and Production Performance in B2MML *(continued)*		
BatchML	*Production Performance (BatchML to B2MML)*	*Production Schedule (B2MML to BatchML)*
DescriptionType:	DescriptionType: SegmentResponse/ Description	DescriptionType: SegmentRequirement/ Description
StatusType:	SegmentStateType: SegmentResponse/ SegmentState	
RecipeIDType:	ProductProductionRuleID: Productionresponse/ ProductProductionRuleID	ProductProductionRuleID: ProductionRequest/ ProductProductionRuleID
BatchIDType:		ProductionRequest/ID: ProductionRequest/ID
LotIDType:	MaterialLotID: Segmentresponse/ materialproducedactual/ MaterialLotID	MaterialLotID: SegmentRequirement/ materialproducedrequirement/ materialLotID
RequestedStartTimeType:		EarliestStartTime: SegmentRequirement/ EarliestStartTime
ActualStartTimeType:	ActualStartTimeType: SegmentResponse/ ActualStartTime	
RequestedEndTimeType:		LatestEndTime: SegmentRequirement/ EarliestEndTime
ActualEndTimeType:	ActualEndTimeType: SegmentResponse/ ActualEndTime	
BatchPriorityType:		PriorityType: ProductionRequest/Priority
RequestedBatchSizeType:	QuantityStringType: MaterialProducedActual/ Quantity	QuantityStringType: MaterialProducedRequirement/ Quantity
UnitOfMeasureType:	UnitOfMeasureType: MaterialActual/Quantity	UnitOfMeasureType: MaterialRequirement/Quantity

Table 5.1. Mapping of the BatchML document to Production Schedule and Production Performance in B2MML *(continued)*		
BatchML	*Production Performance (BatchML to B2MML)*	*Production Schedule (B2MML to BatchML)*
ParameterType:		ParameterType: ProductionParameter/ Parameter
EquipmentIDType:	EquipmentIDType: EquipmentActual/ EquipmentID	EquipmentIDType: EquipmentRequirement/ EquipmentID
EquipmentClassIDType:	EquipmentClassIDType:	EquipmentActual/ EquipmentClassID

third column shows the information that represents the relationship between the information flow from B2MML to BatchML, according to the Production Schedule model. The fields from each document are also identified. Table 5.1 only presents the information elements that would be exchanged, as the batch management software also has information that is generated and managed internally, which is not mapped to other levels, as it is not necessary to transfer all the BatchML information to the MES level or vice versa.

In the Production Schedule document, it is necessary to demonstrate that the existing fields correspond to the MES information and to the information flow from the ERP system and the Batch control system in the exchanges carried out through the B2MML and BatchML documents. This process obtains, as a result, important parameters for a proper ISA-88 and ISA-95 implementation. Regarding the batch control system, the existing fields in the BatchML document correspond to the information flow mapped from the Production Schedule through the middleware. These parameters are necessary for any integration project, regardless of the type of management software. With regards to Production Performance, the information in the B2MML document from Production Performance is not only reported at the end of the production process, but also can be sent to the ERP level during the batch execution. This allows relevant information to include any reactivity level in the integrated system.

Prototype

In order to validate the mapping, a prototype was developed for the middleware that supports the information flow for Production Schedule and Production Performance between the MES and control level using the structure proposed by the

ISA-88 and ISA-95 standards. As shown in Figure 5.2, in the prototype, the Production Schedule is set up according to the business and logistics production system (i.e., ERP) polices and the structure that the B2MML standard uses. Once the Production Performance is set up at the MES level, the software application generates the Production Performance B2MML document, which is sent to the control level through the prototype middleware that transforms the information delivered by the MES system in B2MML to the BatchML document that the control system will use.

Once the control system receives the information form the Production Performance in the BatchML structure, the batch management software sets up the batch process with its information. Once the batch programming is concluded, the batch management software prepares a report of the production in the BatchML document. This XML document is used by the control level to communicate the results of the batch execution to the manufacturing level (i.e., MES); this information then goes through the middleware prototype, which transforms the information in XML delivered by the batch management software using the BatchML structure to the B2MML document, according to the Production Performance structure.

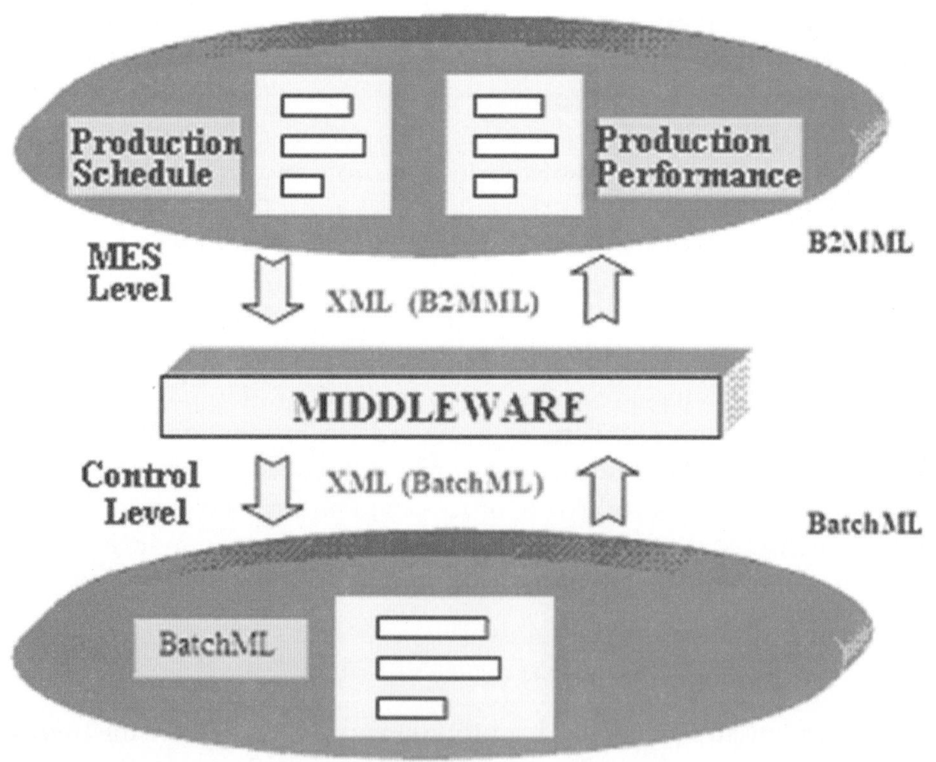

Figure 5.2. Information flow prototype.

Figure 5.3 illustrates the Production Schedule main interface once the corresponding information is introduced. This includes the schedule ID, the equipment ID, the equipment level ID, and the schedule starting and ending time specification.

The Production Performance interface is composed of a production request, the produced material requirements, the consumed material requirements, and the segment requirements. The interface has a control called "save information" that is used to save the production request information once the Production Schedule is set up. Using this command, the B2MML document of the Production Performance in the B2MML structure can be generated. Finally, the interface has a control called "BatchML," which can be used to get to the batch management interface once the production scheduling is carried out. Once selected, the interface shown in Figure 5.4 is presented, which shows the Production Performance mapping corresponding to the BatchListEntry for the packaging stage of a production process of dairy products.

In the same way, Figure 5.5 shows the batch management information mapping in the Production Performance document in B2MML through the middleware prototype. This corresponds to the actual consumed material and the segment response.

Conclusion

Finally, when every interface is completely and correctly set up, it will be possible to store the information and to obtain the Production Schedule, the Production Performance, and the batch control system documents, according to the B2MML and BatchML structure established by the ISA-95 and ISA-88 standards. It will also be possible to store a part of the BatchListEntry document in its BatchML structure, as shown in Figure 5.6.

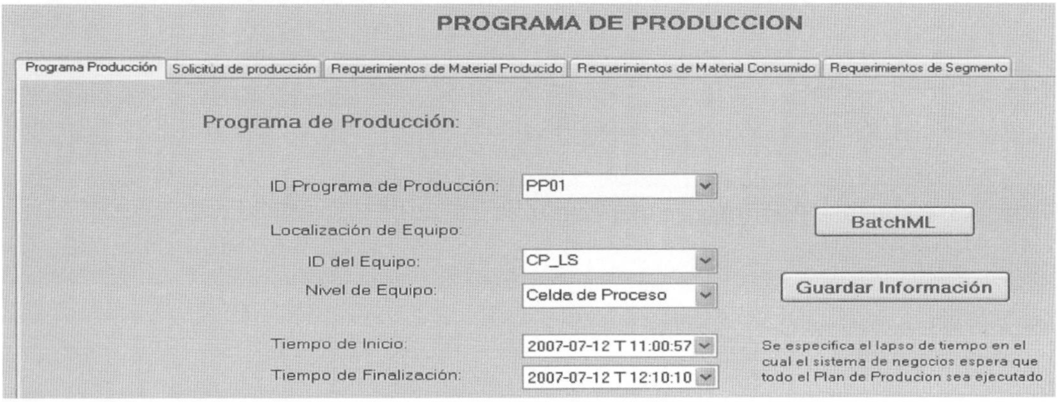

Figure 5.3. Production Schedule interface.

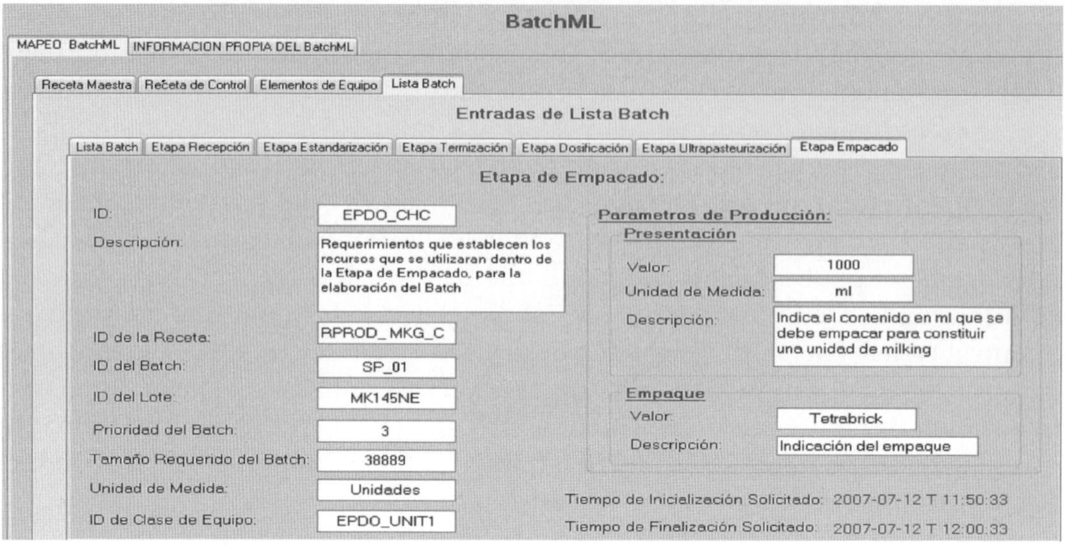

BatchML

MAPEO BatchML | INFORMACION PROPIA DEL BatchML

Receta Maestra | Receta de Control | Elementos de Equipo | Lista Batch

Entradas de Lista Batch

Lista Batch | Etapa Recepción | Etapa Estandarización | Etapa Termización | Etapa Dosificación | Etapa Ultrapasteurización | Etapa Empacado

Etapa de Empacado:

ID: EPDO_CHC

Descripción: Requerimientos que establecen los recursos que se utilizaran dentro de la Etapa de Empacado, para la elaboración del Batch

ID de la Receta: RPROD_ MKG_C

ID del Batch: SP_01

ID del Lote: MK145NE

Prioridad del Batch: 3

Tamaño Requerido del Batch: 38889

Unidad de Medida: Unidades

ID de Clase de Equipo: EPDO_UNIT1

Parametros de Producción:

Presentación

Valor: 1000

Unidad de Medida: ml

Descripción: Indica el contenido en ml que se debe empacar para constituir una unidad de milking

Empaque

Valor: Tetrabrick

Descripción: Indicación del empaque

Tiempo de Inicialización Solicitado: 2007-07-12 T 11:50:33

Tiempo de Finalización Solicitado: 2007-07-12 T 12:00:33

Figure 5.4. Batch list entry mapping.

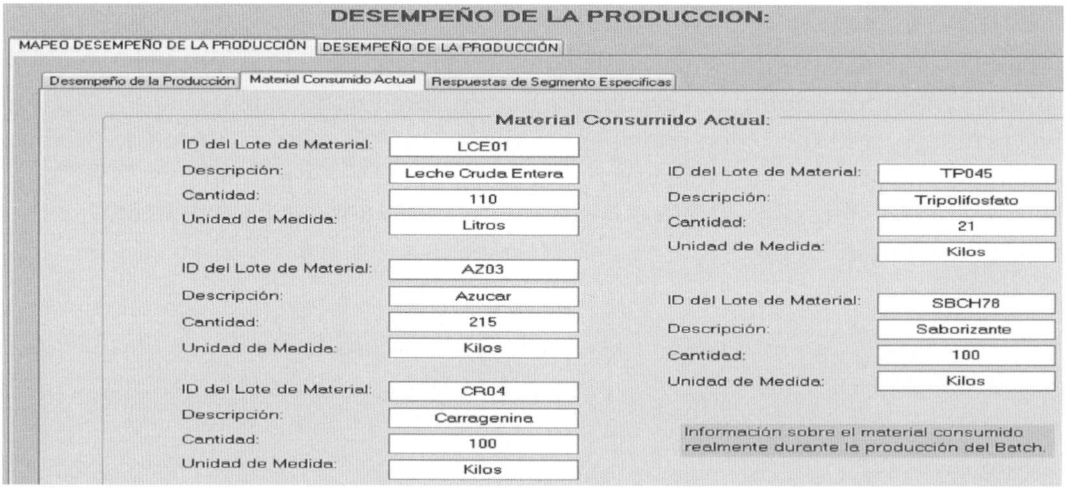

DESEMPEÑO DE LA PRODUCCION:

MAPEO DESEMPEÑO DE LA PRODUCCIÓN | DESEMPEÑO DE LA PRODUCCIÓN

Desempeño de la Producción | Material Consumido Actual | Respuestas de Segmento Especificas

Material Consumido Actual:

ID del Lote de Material: LCE01

Descripción: Leche Cruda Entera

Cantidad: 110

Unidad de Medida: Litros

ID del Lote de Material: AZ03

Descripción: Azucar

Cantidad: 215

Unidad de Medida: Kilos

ID del Lote de Material: CR04

Descripción: Carragenina

Cantidad: 100

Unidad de Medida: Kilos

ID del Lote de Material: TP045

Descripción: Tripolifosfato

Cantidad: 21

Unidad de Medida: Kilos

ID del Lote de Material: SBCH78

Descripción: Saborizante

Cantidad: 100

Unidad de Medida: Kilos

Información sobre el material consumido realmente durante la producción del Batch.

Figure 5.5. Production Performance mapping.

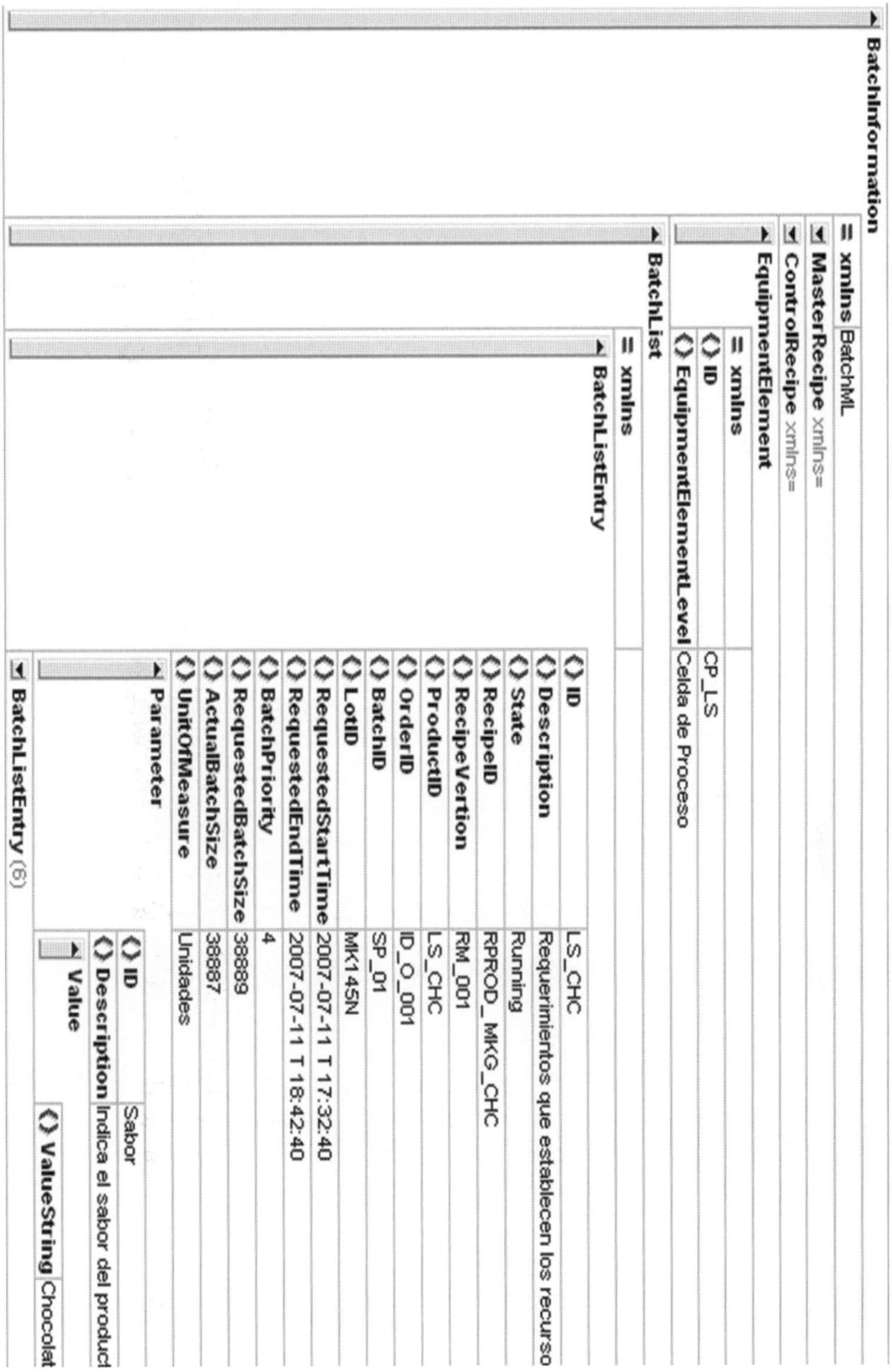

Figure 5.6. BatchListEntry in BatchML.

CHAPTER 6

Implementation of Object Models for a UHT Milk Line as a Case Study

Presented at the WBF
North American Conference,
April 30–May 3, 2007, by

Fabian Yesid Vidal Lopez
Industrial Automation Engineer
fyvidal@unicauca.edu.co
Omnicon Ltd.
Calle 17N No. 9N-20
Cali (Valle del Cauca), 057, Colombia

Libardo Steven Muñoz Trochez
Industrial Automation Engineer
lmunoz@unicauca.edu.co
Omnicon Ltd.
Calle 17N No. 9N-20
Cali (Valle del Cauca), 057, Colombia

Oscar Amaury Rojas Alvarado
Electronic and Telecommunications Engineer
orojas@unicauca.edu.co
University of Cauca
Calle 5 No. 4-50
Popayán (Cauca), 057, Colombia

Abstract

To develop an integrated automation production process, it is necessary to have in mind various factors, such as the process itself, the company's organizational structure, the equipment, and the management of the supplies. The specification and the definition of these factors turns out to be even more complex, due to economic policies and globalization. Thus it is imperative to get the most out of the information flow of a company's business processes, in order to acheive optimal productivity and remain competitive in a globalized market. This article presents a practical vision about how to work with the models established by the ISA-95 standard, which have been widely accepted worldwide to aid in the process of information integration among business and manufacturing systems, thus solving enterprise-control system integration problems.

Introduction

Within companies devoted to the manufacture of food products, a great deal of information is generated that is important for the business systems and the production processes. This information, as a whole, can enable companies to guarantee the fulfillment of the quality standards established by legislation, and in the same way, to increase their productivity levels and to improve the performance in their processes. This translates into great benefits for the business.

To achieve this goal, it is necessary to have effective communication among the different information systems that generate and manage the data in the different levels of the company. This allows the business level to communicate the production requirements and to receive the performance of the process, in order to direct the company policies.

Particularly in the case study analyzed in this chapter, it is essential to integrate the production program information to facilitate the communication of the production requirements, which are established by the business level after analyzing the product demand. These requirements indicate to the Ultra-High Temperature (UHT) milk production area the quantities and the presentations of the dairy product that have to be processed to guarantee the fulfillment of the production requirements. Likewise, it is important for the business level to know the results of the production conducted, detailing the quantities and the types of product, as well as the different resources used, in order to conduct an analysis, determine costs, and establish actions to improve the process.

Due to its background and the great approval that the ISA-95 standard has achieved worldwide, the recommendations and outlines established in the ISA-95 standard were used to solve the difficulties presented by the information exchange between the business level and manufacturing in this case study.

In integration projects where the ISA-95 standard is used, it is necessary to conduct a modeling phase, followed by an implementation phase using middleware to exchange the developed Business To Manufacturing Markup Language

(B2MML) documents. In this case study, efforts have been focused on the information modeling phase, using the ISA-95 standard and the B2MML documents. This is necessary before beginning the development of the implementation phase.

To achieve this goal, in the integration process carried out for the case study, the modeling of the functional structure of the company was carried out using a functional data flow model. Subsequently, object models were used in the analysis and modeling of the material resources, equipment, personnel, and process segments information. Finally, the production and performance program documents in B2MML were elaborated, which allowed the exchange of information in the implementation phase.

Functional Data Flow Model

The development of the functional data flow model has been the most useful tool to understand the functional structure of the company, since it has allowed us to establish in an organized and clear way the operation of the company, detailing in every function the responsible people, the type of decisions they need to make, and the way in which the decisions interact.

In the development of the functional data flow model, an information collection phase was carried out through interviews of the people involved in the activities of quality control, maintenance, inventory, procurement, and production. The questions used in the interview were designed to learn how the activities defined in part 1 of ISA-95 are carried out.

The result of this modeling is a document that clearly identifies which function the different departments of the company, people, or information systems belong to, specifying if the activities that each one of them performs correspond to activities of Level 3 or Level 4. This document has been a great help for the company, since it has allowed it to have more clarity about its business processes and its manufacturing operations. This, in turn, facilitates the identification of the activities that affect the production performance and business process as well, as the information flows between the two and allows for the development of dynamic improvement, control, and optimization procedures. Furthermore, this document has been a great complement to the documentation established in the quality certifications ISO 9000 and 9001.

Advantages of the Functional Data Flow Model

It is important to underline that the functional data flow model was designed in an appropriate and general way, allowing its easy usage and implementation in organizations of all sizes, since all the possible activities and information flows that occur in these companies can be established within the functions and data flows considered in the models. The functional data flow model is not strict concerning its total development, allowing companies to analyze the functionalities and the information flows according to their own requirements.

Object Models

The development of the object models has allowed a bigger approach to the implementation phase, since the obtained information in the analysis of the data flow model is organized and structured within the four general categories defined in the standard—Production Schedule, Product Definition, Production Capability, and Production Performance—which will be interchanged using the enterprise-control interface.

Having in mind the requirements mentioned in this chapter, for the case study, the modeling was carried out for two general information categories that correspond to the Production Schedule and the Production Performance. These models were obtained from the modeling of three resource categories that constitute them: materials, equipment, and process segments. It is necessary to clarify that the information about the personnel resources was not modeled, since it was not considered in the established requirements in the case study.

Although the models are not meant for the designing of databases, they are a great help when defining the main fields in the database that will store the resource information so that it facilitates the information exchange in the way proposed by ISA-95.

Materials Model

Using the materials model, a description of the materials that are used in the production process was conducted, which included information about the raw materials, medium products, byproducts, and finished products.

Bearing in mind the goals of the project and the type of information necessary to build the production program and the production performance report in the B2MML structure, the instances of the materials model have only included the object definitions of the material class and material definition with their own properties. The information of the material lot and sub-lot has not been included since, for the case study, this component has already been defined, while the material lot to use is specified in the documents of the production program and production performance. To produce an instance of the materials model, the following steps were carried out:

1. Identify the material resources that are considered raw material, intermediate products, and finished products

2. Define the IDs and properties of each one of the materials

3. Group the defined materials in general classes

4. Develop the material definitions and the material classes using the B2MML structure

For the development of steps 1, 2, and 3, the documents previously developed for ISO 9001 certification were taken into account, which facilitated the fulfillment of the objectives proposed in this phase of the project. Additionally, in order to facilitate development of the B2MML material model a document was produced that included the required information. Figure 6.1 presents the B2MML document for the materials object model.

Equipment Model

The information concerning all the equipment involved in the manufacturing process of the UHT flavored milk was organized and structured using the equipment model. To conduct the equipment model instance, the following steps were followed:

1. Make reference to the equipment hierarchy model previously developed, in order to identify and organize the equipment that is used in the manufacturing of flavored milk

2. Define the equipment classes and their properties, grouping them according to similar characteristics

3. Define the equipment with their IDs and properties

4. Develop the equipment class and the equipment with their properties in the B2MML structure

After conducting an analysis based on the equipment hierarchy model, it was identified that in the flavored milk manufacturing process, a process cell is composed of six units. Figure 6.2 shows the definition of an equipment class in the B2MML structure for a heating unit.

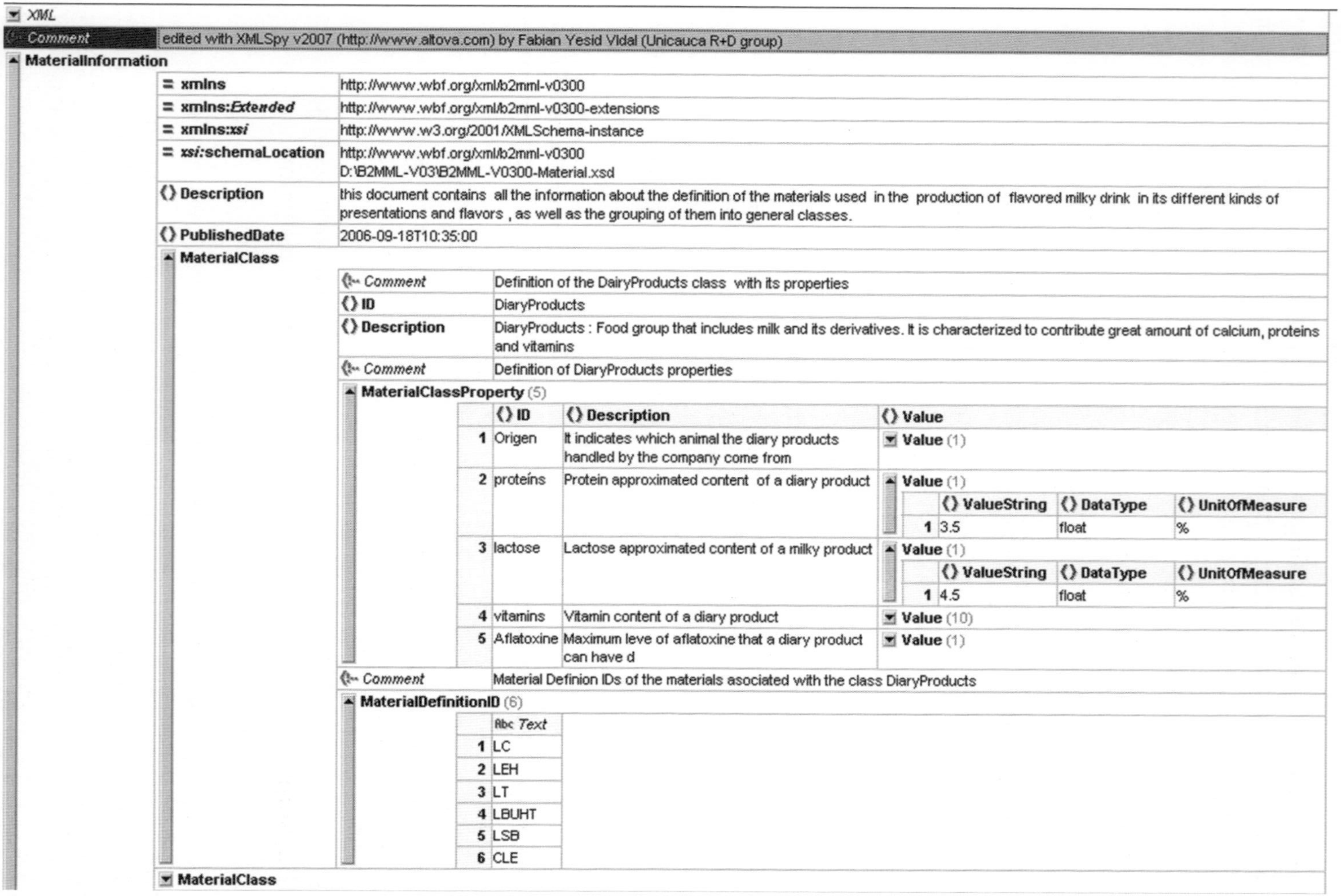

XML

| Comment | edited with XMLSpy v2007 (http://www.altova.com) by Fabian Yesid Vidal (Unicauca R+D group) |

MaterialInformation

= xmlns	http://www.wbf.org/xml/b2mml-v0300
= xmlns:*Extended*	http://www.wbf.org/xml/b2mml-v0300-extensions
= xmlns:*xsi*	http://www.w3.org/2001/XMLSchema-instance
= *xsi*:schemaLocation	http://www.wbf.org/xml/b2mml-v0300 D:\B2MML-V03\B2MML-V0300-Material.xsd
() Description	this document contains all the information about the definition of the materials used in the production of flavored milky drink in its different kinds of presentations and flavors , as well as the grouping of them into general classes.
() PublishedDate	2006-09-18T10:35:00

MaterialClass

Comment	Definition of the DairyProducts class with its properties
() ID	DairyProducts
() Description	DiaryProducts : Food group that includes milk and its derivatives. It is characterized to contribute great amount of calcium, proteins and vitamins
Comment	Definition of DiaryProducts properties

MaterialClassProperty (5)

	() ID	() Description	() Value			
1	Origen	It indicates which animal the diary products handled by the company come from	Value (1)			
2	proteins	Protein approximated content of a diary product	Value (1)			
				() ValueString	() DataType	() UnitOfMeasure
			1	3.5	float	%
3	lactose	Lactose approximated content of a milky product	Value (1)			
				() ValueString	() DataType	() UnitOfMeasure
			1	4.5	float	%
4	vitamins	Vitamin content of a diary product	Value (10)			
5	Aflatoxine	Maximum leve of aflatoxine that a diary product can have d	Value (1)			

| Comment | Material Definion IDs of the materials asociated with the class DiaryProducts |

MaterialDefinitionID (6)

	Abc Text
1	LC
2	LEH
3	LT
4	LBUHT
5	LSB
6	CLE

MaterialClass

Figure 6.1. Materials model stage in B2MML.

Equipment			
Comment	PACKAGING UNIT		
Equipment			
Comment	************************************* CLASS DEFINITIONS ***********************************		
Comment	RECEPTION UNIT CLASS		
EquipmentClass			
Comment	STANDARIZATION UNIT CLASS		
EquipmentClass			
Comment	HEATING UNIT CLASS		
EquipmentClass			
	ID	TZN_UNIT	
	Description	Heating Unit Class: This class collects equipment with similar features to the heating units	
	EquipmentClassProperty		
		ID	CAP_TZN_UNIT
		Description	Capacity of the Heating Unit: processing capacity of the unit
		Value	
			ValueString
			DataType — nonNegativeInteger
			UnitOfMeasure — Liter/Hour
	EquipmentID	TZN_UNIT1	
Comment	ADDITION OF ADDITIVES UNIT CLASS		
EquipmentClass			
Comment	ULTRA HIGH TEMPERATURE HEATING UNIT CLASS		
EquipmentClass			

Figure 6.2. Definition of an equipment class in the B2MML structure.

Process Segment Model

After conducting an analysis of the ISA-95 standard, we were able to conclude that the definition of a process segment includes the identification of all the resources used in a common way in a production step for the manufacturing of different products. This does not include the resource specifications that are required for a product in particular, since this information is detailed in the product segments. Therefore, the process segment model has allowed us to establish a general view of the manufacturing process for the business system, letting us know the activities that the manufacturing system is able to perform and the materials, equipment, and personnel used in each of the production steps.

Having in mind the production system of flavored milk established in the study case, six segments within the process cell for the flavored milk were identified, which allow carrying out the reception, heating, addition of additives, standardization, UHT heating, and packaging activities. Each one of these activities constitutes a process segment in a global segment that represents the resources in the process cell.

Due to the fact that the process segments can be considered as a logical view of the production operation, such segments were established in a way that facilitated the Production Schedule and the Production Performance documents, according to the requirements of the case study.

When considering the process cell as a process segment, versatility at the time of conducting the production scheduling was obtained, indicating to a global "being" (e.g., process cell), in a general way, the kind of product to be elaborated, the quantity, the duration for the production requests, and the resources that needed to be used. In this case study, this general information cannot be obtained when doing the specification of each one of the process segments that constitute the process cell. For the development of this process segments model, the following steps were followed:

1. Identify the process segments

2. Identify the existing dependencies between the process segments

3. Establish the routing for the process segments

4. Specify the material and equipment segments for each of the process segments

5. Specify the process segments identified in the B2MML structure

Figure 6.3 shows the B2MML document corresponding to the process segment model.

Production Schedule Model

Having in mind the requirements of the case study and the definitions of the material resource models, equipment, and process segments obtained in the B2MML, certain parameters are communicated to the flavored milk process cell to define a schedule for a product. They included the type of product and the quantity that has to be produced, the quantity of raw milk to be used, the presentation (e.g., milliliters content in the package), the ID lot, the date when the production request has to be finished, and so on.

For the segments that compose the flavored milk process segment, the segment requirements are established by specifying the equipment and material requirements, as shown in Figure 6.4.

For the specification of the material requirements to be used in each process or product segment, the information defined in the product segments that corresponds to the instance of the product model definition was taken into account. Thus it can be concluded that materials information is only differentiated by the

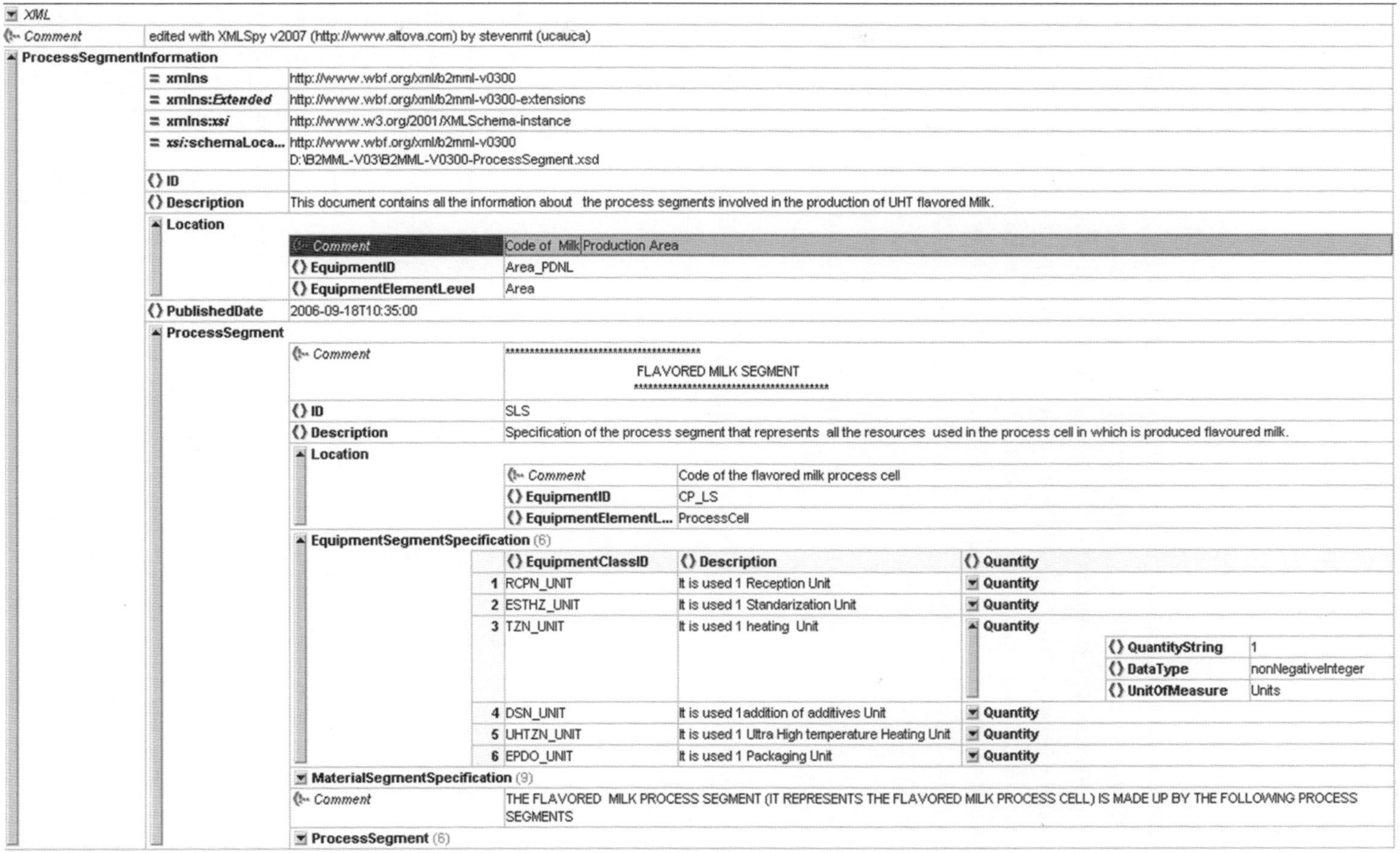

The figure is a screenshot of an XML editor tree view:

▼ XML			
← Comment	edited with XMLSpy v2007 (http://www.altova.com) by stevenmt (ucauca)		
▲ ProcessSegmentInformation			
≡ xmlns	http://www.wbf.org/xml/b2mml-v0300		
≡ xmlns:Extended	http://www.wbf.org/xml/b2mml-v0300-extensions		
≡ xmlns:xsi	http://www.w3.org/2001/XMLSchema-instance		
≡ xsi:schemaLoca...	http://www.wbf.org/xml/b2mml-v0300 D:\B2MML-V03\B2MML-V0300-ProcessSegment.xsd		
◇ ID			
◇ Description	This document contains all the information about the process segments involved in the production of UHT flavored Milk.		
▲ Location			
← Comment	Code of Milk Production Area		
◇ EquipmentID	Area_PDNL		
◇ EquipmentElementLevel	Area		
◇ PublishedDate	2006-09-18T10:35:00		
▲ ProcessSegment			
← Comment	************************************ FLAVORED MILK SEGMENT ************************************		
◇ ID	SLS		
◇ Description	Specification of the process segment that represents all the resources used in the process cell in which is produced flavoured milk.		
▲ Location			
← Comment	Code of the flavored milk process cell		
◇ EquipmentID	CP_LS		
◇ EquipmentElementL...	ProcessCell		

EquipmentSegmentSpecification (6)

	◇ EquipmentClassID	◇ Description	◇ Quantity	
1	RCPN_UNIT	It is used 1 Reception Unit	▼ Quantity	
2	ESTHZ_UNIT	It is used 1 Standarization Unit	▼ Quantity	
3	TZN_UNIT	It is used 1 heating Unit	▲ Quantity	◇ QuantityString 1 / ◇ DataType nonNegativeInteger / ◇ UnitOfMeasure Units
4	DSN_UNIT	It is used 1addition of additives Unit	▼ Quantity	
5	UHTZN_UNIT	It is used 1 Ultra High temperature Heating Unit	▼ Quantity	
6	EPDO_UNIT	It is used 1 Packaging Unit	▼ Quantity	

▼ MaterialSegmentSpecification (9)		
← Comment	THE FLAVORED MILK PROCESS SEGMENT (IT REPRESENTS THE FLAVORED MILK PROCESS CELL) IS MADE UP BY THE FOLLOWING PROCESS SEGMENTS	
▼ ProcessSegment (6)		

Figure 6.3. Instance of the process segment model in B2MML.

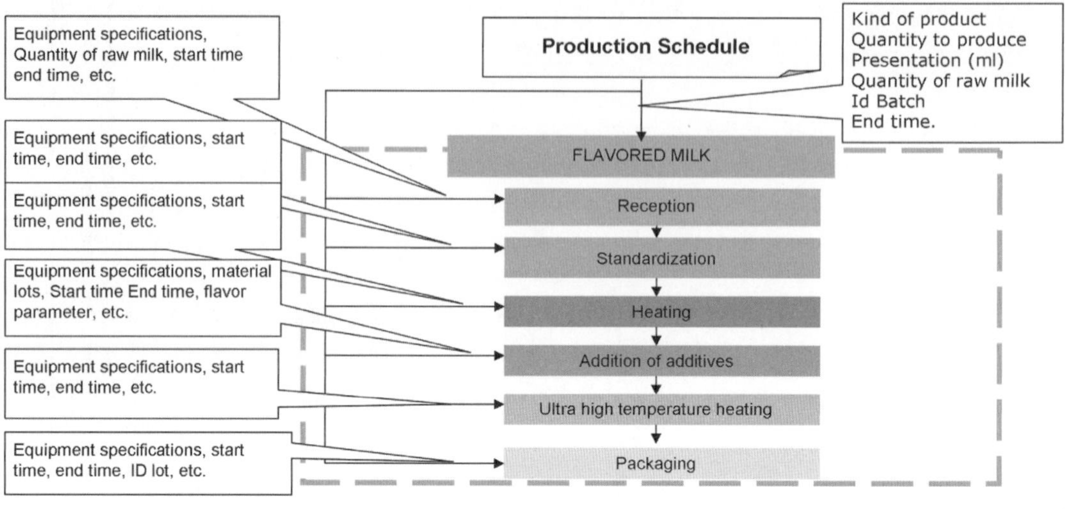

Figure 6.4. Process segment scheduling carried out by the production schedule, which requests the execution of a specific batch of flavored milk.

material lot specification that is defined in the production schedule. The specification of the equipment requirements was conducted using the equipment ID class and not directly with the equipment ID, in order to have flexibility when assigning the equipment.

It is important to mention that in the production schedule, apart from indicating to Level 3 the production requirements through the information contained in the "segment requirement" field, Level 4 can also request important information to be returned at the end of the production execution using the information contained in the "segment response" field. The following questions arise here: Are the existing MES systems intelligent enough to identify the information to be returned? Or is it necessary to use an external application to carry out the identification of this information? For the case study, the production schedule only specifies the production requirements, and the required information is reported with a defined structure at the end of the production execution, which cannot be modified in a dynamic way using the Production Performance document.

Production Performance Model

Using the Production Performance document instantiated in B2MML, Level 4 can know the real data of what happens in the process, including the recipe used and the generated product and its quantities, as well as the important parameters in the production process, the materials, the equipment, and their operation times.

The B2MML document of the Production Performance that has been obtained by processing the instance of the respective object model allows Level 3 to report the parameters and their respective values of what happened in the production process, using two detail levels. The first one provides, through the response of the flavored milk process segment, data for the order execution times, the quantities, and the consumed material lot, as well as the quantities and the material lot produced. In addition, through the segment responses of the segments that compose the flavored milk process, a high detail level was reached, obtaining the specification of the process stages that the materials and equipment used, along with the intermediate products generated. Figure 6.5 shows the B2MML document corresponding to the production schedule.

ProductionResponse

⇜ Comment	Production Response ID
⟨⟩ ID	RP01
⇜ Comment	This is the production request for what it is being specified the performance
⟨⟩ ProductionRequestID	SP01
⇜ Comment	Production rule that was used for the production request execution
⟨⟩ ProductProductionRuleID	RPROD_ MKG_CHC
⇜ Comment	actual time of the production request execution
⟨⟩ StartTime	2007-01-25T08:15:00
⟨⟩ EndTime	2007-01-25T15:58:00
⇜ Comment	Performance Specification by segments

SegmentResponse

⟨⟩ ID	SLS
⟨⟩ ProductSegmentID	LS_CHC
⟨⟩ Description	Report about the actual resources used in the flavored milk process cell to execute the production request
⟨⟩ ActualStartTime	2007-01-25T08:15:00
⟨⟩ ActualEndTime	2007-01-25T15:53:00

ProductionData (2)

EquipmentActual

⟨⟩ EquipmentID	CP_LS
⟨⟩ Description	flavored milk process cell

Location

⟨⟩ EquipmentID	Area_PDNL
⟨⟩ EquipmentElem...	Area

Quantity

MaterialProducedActual

⟨⟩ MaterialDefinitionID	MKCHC
⟨⟩ MaterialLotID	MK145N
⟨⟩ Description	chocolate flavored milk

Quantity

MaterialConsumedActual (8)

	⟨⟩ MaterialDefinitionID	⟨⟩ MaterialLotID	⟨⟩ Description	⟨⟩ Quantity
1	LCE	LCE01	Raw milk	Quantity
2	CLE			Quantity
3	AZ	AZ03	Sugar	Quantity
4	CRRG	CR04	carragenin	Quantity
5	TRPF	TP045	Tripolifosfate	Quantity
6	SBCHC	SBCH78	chocolate flavor	Quantity
7	SBVAIN	SBVN15	Vanilla flavor	Quantity
8	TBK	TB045	tetrabrick	Quantity

⟨⟩ QuantityString	6910
⟨⟩ DataType	nonNegativeInteger
UnitOfMeasure	Liters

SegmentResponse (6)

Figure 6.5. Instance of the Production Schedule model in B2MML.

B2MML for Plant Maintenance and Asset Management in a Global, Service-oriented MES Scenario

Presented at the WBF
European Conference,
November 10–12, 2008, by

Frank Bezjak, Dipl. Ing.
frank.bezjak@bayertechnology.com
Bayer Technology Services GmbH
Kaiser-Wilhelm-Allee 7
Leverkusen, D-51368, Germany

Martin Zeller, Dipl. Ing.
martin.zeller@bayertechnology.com
Bayer Technology Services GmbH
Kaiser-Wilhelm-Allee 7
Leverkusen, D-51368, Germany

Abstract

Plant Maintenance and Asset Management are two key factors for effective management of production equipment in the chemical and pharmaceutical industry. ISA-95 part 3 describes models for maintenance processes, among others. These may differ from the process in real use, and Business To Machine Markup Language (B2MML) schemata may not completely fit a particular situation.

This chapter shows how to meet the challenge to integrate a Manufacturing Execution System (MES) with a SAP Enterprise Resource Planning (ERP) system for maintaining plant assets with the use of B2MML to exchange data, even though the process is not ISA-95 compliant. The scenario involves plant assets from various plants that are stored in a customized SAP Enterprise Core Component (ECC) Plant Maintenance (PM) module. The focus is on specifying, designing, extending, and applying ISA-95 and B2MML to non-production operations through the use of Service-Oriented Architecture (SOA).

The solution described in this chapter shows how the B2MML data model may be extended with a representation of the customized SAP business object to fit the company's landscape. The integration of vendor-specific data models that are involved to match the described scenario required another solution. Yet another solution is given for keeping a representation of existing plant assets, including structures and elements, as unique data objects throughout the entire company and how to communicate them via SOA.

Introduction

Safety is one the most important aspects in the chemical and pharmaceutical industry, since many of the raw materials, intermediates, and products that are handled are hazardous. Continuous or routine maintenance of production equipment is a basis for safety in a production environment, and therefore it is the focus of Plant Maintenance.

Yet from the economic side of things, it is very important that operator and machine downtime is reduced to a minimum. Hence the management of maintenance processes has to be as efficient as possible in order to meet both requirements. With better data, the management of maintenance is improved, and safety is increased.

The situation today is that most of the various MES at this particular company do not have a direct link to ERP or offer a proprietary solution to connect the systems via SAP's proprietary communication. First of all, this means that the data have to be exchanged manually with the use of the SAP user interface. Secondly, an MES vendor-specific integration solution may be complex and therefore cost intensive, depending on the technical skills of its IT experts.

With the application of automated data collection techniques, on the basis of service-orientated architecture and the mapping of business and production data, the maintenance process can be done more efficiently today.

The question for a global operating company is, how far does this concept fit into the already existing IT environment, and how cost intensive is the application?

Consider the fact that the ERP system inside the company is already customized and specialized for several business purposes. Business experts in the company recognized the need to make both of these processes more efficient. Therefore, the company initiative for research projects initiated a project to enhance the introduced processes.

The task of the project was to integrate MES and ERP with SOA and B2MML at a pharmaceutical plant. Therefore, the following scenario takes place in a regulated Good Manufacturing Process (GMP) environment. The data that are communicated between the systems should be partially automated and partially controllable via a Web interface.

Because Asset Management and Plant Maintenance are very important topics at every Bayer production site and plant around the world, this solution should be an option for more than one plant. However, another side of the coin is that the work required for the data coupling and the customization of MES and SAP has to be reduced to a minimum.

The Business Process

At Bayer, the ERP system holds a representation of the various plant assets, such as production equipment and the corresponding equipment status, in an SAP ECC PM module. This module is called Technical Shift Book (TSB), a development by Bayer Business Services (BBS).

All events that occur during a production shift are documented and recorded in the TSB. Each of these events is a message, which is directly related to the existing plant assets, including structures and elements. The data objects are unique in the whole IT structure of a company. To process a message successfully, certain data are mandatory. This information is either stored in historian databases or is available in an MES system.

At Bayer, every equipment or logical plant structure can be identified according to a proprietary plant identification code. This identification code, which is also called AKZ, differs in its structure from standards like ISA-88.

Most MES systems use a proprietary code to map the existing plant structure. The main problem that arises from this fact is that an MES needs to know these unique IDs in order to complete the necessary data for the SAP messages.

Figure 7.1 shows how the messages are processed in the TSB module. In this situation, the operator enters the messages manually into the SAP PM module. He completes the necessary data field with the information from the MES system.

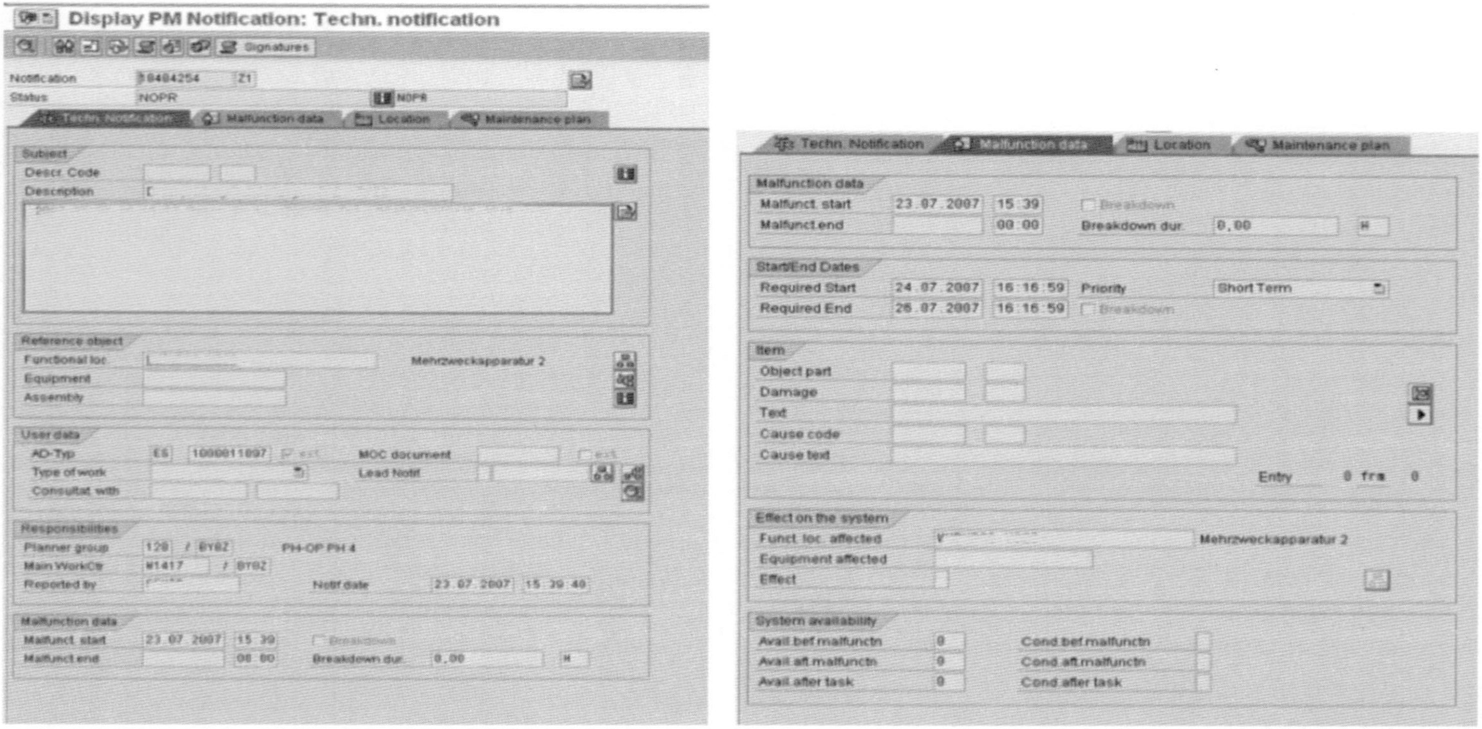

Figure 7.1. TSB for Asset Management.

Currently, there are four types of messages that are processed and documented in the SAP system, according to the TSB policy. These messages were identified as relevant for the integration process:

- Z1 messages are related to technical interferences. These events occur if some process steps or operations fail during the production process. The result of this failure does not have to lead to machine downtime or bad production.

- Z2 messages document events that occur when regular maintenance work takes place on a specific plant unit without interfering with the actual production process. Furthermore, these messages can be related directly to the Z1 messages because equipment has to be checked if an anomaly occurs.

- Z3 messages document any other events that do not relate to the events connected with the Z1 or Z2 messages. This message type is useful for documenting extraordinary information.

- Z4 messages can be the start, the end, or the total duration of a batch. Furthermore, they may be the downtime of a machine or the cleaning interval.

Technical Concept

The main task of the integration was to automate the completion of the data field through the use of the MES data. The data were divided into two groups: The first group contained the Z1 to Z3 messages. This group needed to be viewed before being communicated to SAP. The second group, the Z4 messages, could be communicated to SAP without further processing. The MES can identify the process in SAP with the resulting SAP Message ID.

B2MML has been recognized as the chance to bridge the gap with the intermediate Extensible Markup Language (XML) data format to integrate the systems and to harmonize the different standards. Beside the fact that MES systems vary from production site to production site, there are several MES that do not offer a B2MML interface or even an interface for XML-based communication. Therefore, the integration of future scenarios can be rather complex.

Though the business data are available and the MES can complete the information necessary for the messages, the solution uses additional data sources. Therefore, the availability and the flexibility are given to meet the key requirements for a global scenario.

The main idea is to design a software solution that handles the data communication with SAP XI and offers various possibilities to integrate existing MES systems into the company landscape. This software should be installed and integrated into the IT environment of the plant.

Due to the fact that many IT experts at Bayer have .NET knowledge, a Microsoft server solution is used as the platform for this integration. This makes installation and customization very handy for the experts worldwide.

Infrastructure

During the integration of MES and SAP systems, the MES layer is the most critical part of the integration, due to the fact that the integration has to be done in two directions. One part of the implementation is done with SAP, and the other part is done with the process control layer. Therefore, different data models need to be integrated.

Figure 7.2 illustrates the realized scenario with the SAP NetWeaver that hosts the ECC PM module and the middleware XI. Besides the ERP, which is hosted at the BBS headquarters, the additional server component and the MES are shown on the plant site.

Integration of the B2MML Maintenance Schema into SAP XI

This section describes the mapping of the B2MML maintenance schema. The result of this work, which was done in SAP XI, is a Web service that can create messages in SAP and then return the unique SAP message number, which is used for monitoring purposes.

The basic idea is that Z1 through Z4 messages can also be generated with a Business Application Programming Interface (Bapi) method, which is based on SAP Remote Function Call (RFC) communication. Therefore, the first step of the integration was the mapping of the Bapi to the B2MML-V0300-material schema. The Bapi requires the following mandatory data fields:

- *FUNCT_LOC*. The SAP ID that identifies the target equipment according to the AKZ structure.

- *SHORT_TEXT*. The main cause for the event.

- *NOTIF_TYPE*. The type of message (i.e., Z1 to Z4).

The most important question was how the data could be mapped to the required data fields. Besides the fact that this solution was implemented in a

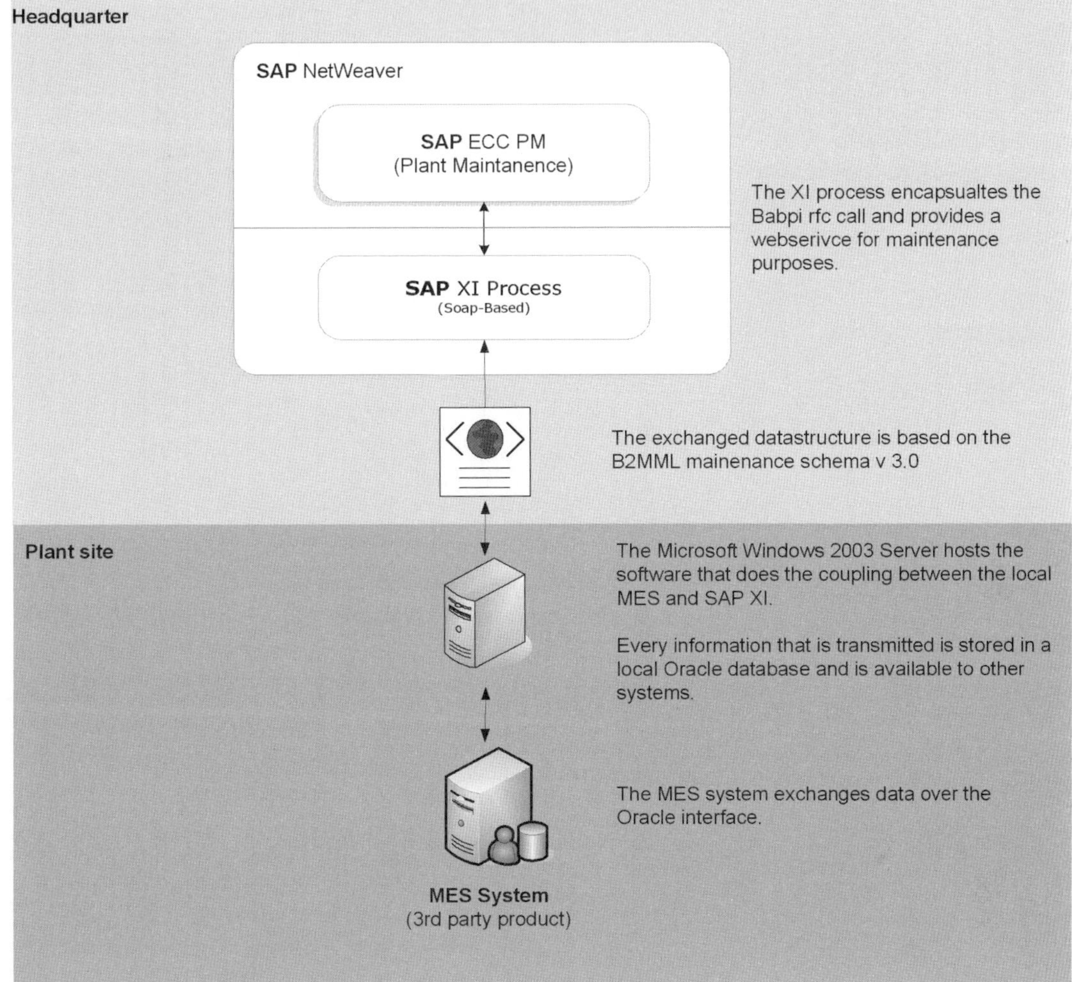

Figure 7.2. The whole system landscape.

pharmaceutical plant, future scenarios in the chemical and pharmaceutical plants across the world should also be included in development efforts.

The required B2MML mapping involved just a few data fields. Because the TSB module is highly customized, the standard B2MML syntax does not provide the necessary structure to fulfill a complete, meaningful mapping between the Bapi and the Web service. Therefore, the necessary B2MML structure was stripped down to a minimum.

Table 7.1 illustrates a message form based on the ISA-95 Maintenance Information schema. The mandatory Bapi fields are encapsulated in the XML structure

Table 7.1. The final B2MML maintenance schema

```xml
<?xml version="1.0" encoding="utf-8"?>
<xs:schema xmlns:xsi="http://www.w3.org/2001/XMLSchema-instance"
attributeFormDefault="unqualified" elementFormDefault="qualified"
targetNamespace="http://www.wbf.org/xml/b2mml-v0300"
xmlns:xs="http://www.w3.org/2001/XMLSchema">
 <xs:element name="MaintenanceInformation">
   <xs:complexType>
      <xs:sequence>
         <xs:element name="ID" type="xs:string">NOTIF_TYPE</xs:element>
         <xs:element name="Description" type="xs:string">SHORT_TEXT</xs:element>
         <xs:element name="MaintenanceRequest">
            <xs:complexType>
               <xs:sequence>
                  <xs:element name="EquipmentID"
                  type="xs:string">FUNT_LOC</xs:element>
               </xs:sequence>
            </xs:complexType>
         </xs:element>
      </xs:sequence>
 </xs:complexType>
</xs:element>
</xs:schema>
```

and highlighted. The outgoing message is a Maintenance Request and the incoming message is a Maintenance Response. Both are encapsulated in the Maintenance Information. Finally, this process is offered through Web Services Description Language (WSDL), and this is the basis for the Coupling Software on the plant site.

Application of the B2MML Proxy and Data Coupling

After the Web service was built in SAP XI, the concepts of the Coupling Software Component (CSC) on the production site were illustrated. The purpose of this was to integrate the MES system via a database coupling, create a B2MML proxy toward SAP XI, and manage outgoing and incoming messages.

For this purpose, a server was installed with a Microsoft .NET Framework, Microsoft Internet Information Services (IIS), B2MML schemas, and an Oracle database. With the use of the .NET Framework, a Web service proxy was generated and deployed in the IIS environment. Through this proxy, the communication to SAP XI was possible. The integrated software layers are shown in Figure 7.3.

The plant structure, according to SAP, was stored in the Oracle database of the plant software component. This part of the system is represented as the "Coupling Server Database" in Figure 7.3. The business data that are exchanged between the B2MML coupling component and the MES are stored in translation tables. Because both systems use an Oracle 10.2 database, it was relatively easy to establish the communication.

Communication of the Z1 to Z3 Messages

In order to communicate and to view the message types Z1 to Z3, an ASP.NET Web site was programmed. This site contains a grid that shows the data available in the transfer tables of the Oracle database. The user is able to select certain messages and send them to SAP.

After the Message ID is returned from SAP, it is stored in the database and viewed in the data grid, so the operator is able to trace a message. The main benefit that arises from the Web interface is that no additional software is needed on the operator's Human Machine Interface (HMI).

Automation of the Z4 Messages

To automate the communication of the Z4 messages, a Windows Service was implemented on the Microsoft server. If updates occur on the transaction tables, then the Windows Service is triggered and starts to invoke a Web service call. After the information is processed, the message ID is stored in the database.

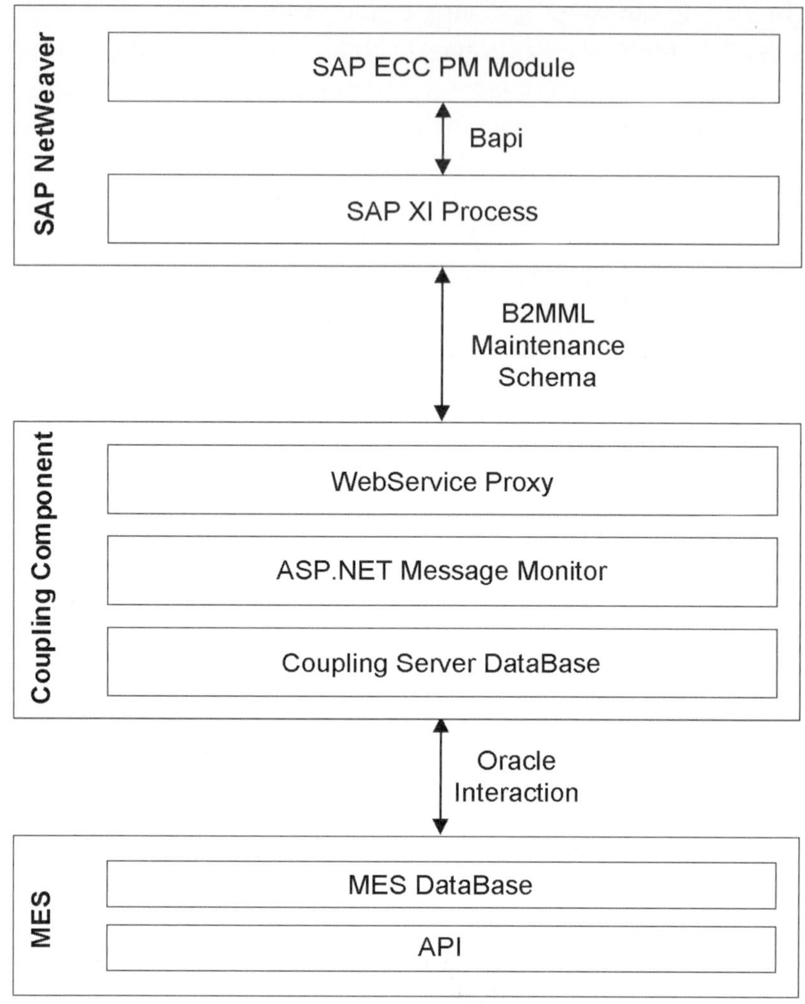

Figure 7.3. System setup.

Conclusion and Outlook

The project using B2MML to exchange data on Plant Maintenance between MES and ERP was carried out successfully. It was essential that an interface based on standards was established, although proprietary formats are used on the ERP and MES sides.

Using SAP XI on the ERP side, a Web server (Proxy) on the MES side, and B2MML-based schemata in between offers maximum reusability in the future and an out-of-the-box solution that is ready for a global integration scenario.

Using B2MML delivers an advantage from a purely technical perspective. In our experience, the advantage from an organizational perspective is even more important. Application specialists usually are familiar with either ERP or MES. For integration purposes, both groups developed a joint solution.

For each data element, exchanged syntax and semantics have to be clarified. After both groups learned to use B2MML, it was rather easy to run the integration, since both groups had a common understanding based on the syntax and semantics of B2MML. Thus the technical interface also serves as an organizational interface.

In the future, systems with built-in B2MML functionality will be available. Additional interface layers may then become obsolete. The usage of B2MML will then be even more advantageous. In the pilot project, B2MML was used for Plant Maintenance. Currently, B2MML is being used in projects that are focused on production. These projects also benefit from the described advantages.

CHAPTER 8

Data Transaction Efficiency in Batch Control: Eliminating the Middleman

Presented at the WBF
North American Conference,
April 30–May 3, 2007, by

Scott W. Sommer, P.E., C.A.P.
Automation Technology Manager
scott.sommer@jacobs.com
Jacobs Engineering Group
2 Ash Street
Suite 3000
Conshohocken, PA 19460, USA

Abstract

A new paradigm is emerging in which the "middleman" in batch data transactions is eliminated, thereby eliminating the potential errors and timing issues that occur as data pass through intermediate applications, gateways, and protocol conversions. This chapter will explore methods of moving data directly from the source to the final point of use—whether this is an operator console, a controller, or data archived on disk media—without passing through these intermediate databases, applications, and other "middlemen." By design and through careful construction of the requirements documents, data producers will communicate directly with data consumers. This approach is described as the Directed Data Transaction Model (DDTM) in this chapter.

The DDTM efficiently splits control applications between machine-level control requirements, cell-level control requirements, and enterprise control requirements. Rather than following the current approach of making data available to all control elements at all control levels, Directed Data Transaction (DDT) forms requirements based on data producers and data consumers and structures the design of the control system, including hardware and application software, to meet these stated requirements.

Introduction

ANSI/ISA-95.00.03-2005[2] defines the multi-level functional hierarchy of activities as being composed of five levels, labeled Level 0 through Level 4. Each level encompasses a time domain (e.g., msec, sec, min, hours, shifts), an activity domain (e.g., scheduling, production, monitoring), and a data domain (e.g., inventory levels, production records, sensor values).

In modern control systems, application software has been written to address the particular needs of a particular level in the hierarchy with minimal overlap. For example, Level 1 applications are typically encoded in firmware or proprietary hardware application spaces, such as in a Programmable Logic Controller (PLC) or a Distributed Control System (DCS). Here the control engineer creates a program or configures blocks and program elements and then downloads the program or configuration to the controller.

At the Level 2 application space, Supervisory Control And Data Acquisition (SCADA) and DCS vendors have produced a vast array of applications that reside on commercially available computer hardware as client-server applications. Graphics, applications, and data services normally reside on a server-class computer, with many client workstations located at the end-use points on the plant floor. Network connections link the Level 1 and Level 2 application components together, thereby forming a "web" of nodes where data and applications can share resources, data values, and information.

At the Level 3 domain, data from the Level 1 and Level 2 systems are shared with a Manufacturing Operations Management (MOM) application. This application is used to set up and control workflows, perform detailed production scheduling, and maintain data records to be used for reporting and production optimization. The Level 3 applications typically reside across a separate bounded network, since the data passed between this level and the Level 1 and Level 2 applications are typically transactional, as opposed to scheduled.

In theory, this tightly integrated network of data and applications performs in an industrial environment with a high reliability (>99.99%) and minimal failures or required operator or maintenance intervention. Even with this high reliability, an hour of downtime per year due to network hardware failures alone can be expected.

In one recent poll, 8% of respondents indicated they experienced network issues weekly, and an additional 11% indicated rising numbers of incidents on their network.[3] A typical batch control system architecture is shown in Figure 8.1.

Data Transactions in Batch Control Systems

The batch control system architecture depicted in Figure 8.1 appears to be quite efficient, with process cell–based functions handled by the lower "island of automation" (process cell 1) and the common server functions handled by the upper "island." However, there are problems, even in this simple case.

While all systems manufacturers claim to provide a reliable and efficient architecture for data transactions, the reality is that most control systems' reliability is far less than optimal due to many factors,[3] including the following:

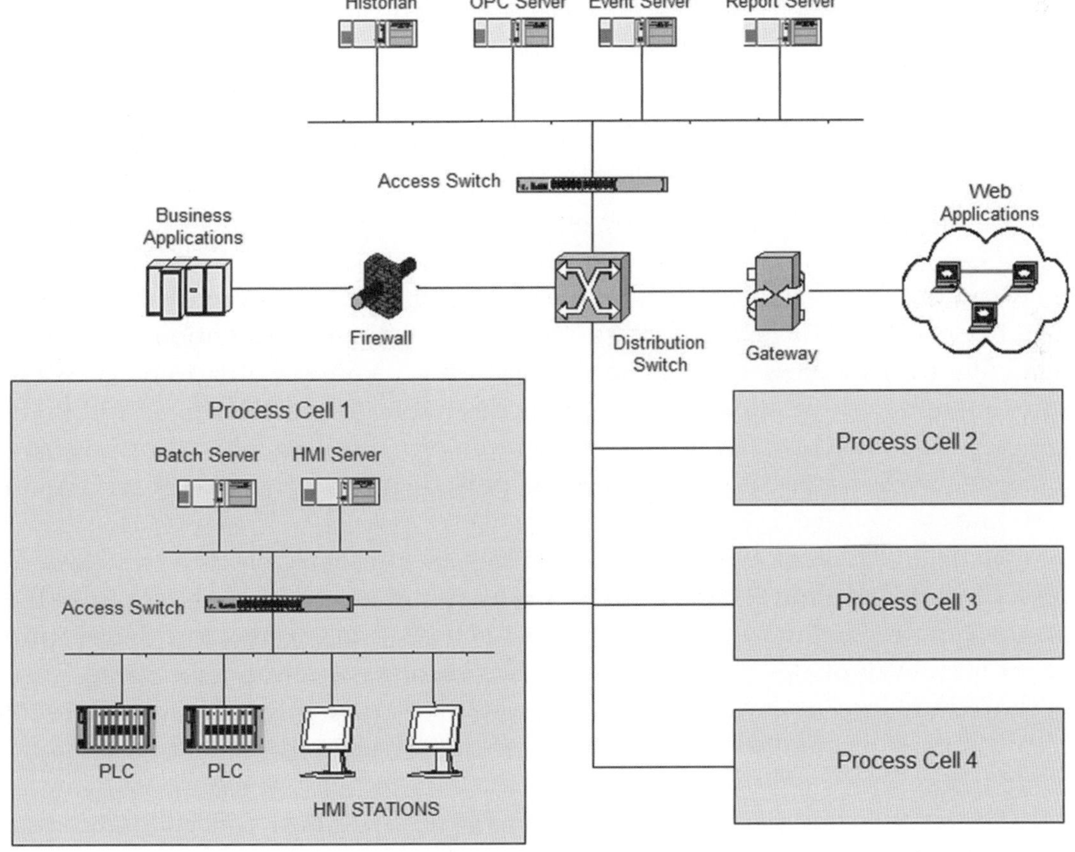

Figure 8.1. Simplified modern batch control network architecture.

- Applications from multiple vendors that do not always "play well" together

- Network loading issues due to poor network design, "data bursting," or "data flooding"

- Operating system applications errors

- Hardware incompatibilities, device driver limitations, and so on

- "Fragile" communications links between applications (e.g., Distributed Component Object Model [DCOM])

- Limitations of application software (e.g., too many transactions, limits on memory or disk space)

- Applications that become "out of sync" or require intervention to continue

- Memory leaks, incompatible Dynamic-Link Library (DLL) files, or internal software anomalies

- Applications that are pushed beyond their intended limitations and use

- Changes made to applications, configurations, or operating systems that have not been fully tested

- Internal application software errors

Vendors, systems integrators, and end users alike are working daily to eliminate these issues. However, the issues listed will remain as limitations to system reliability over the next 10 years because (1) vendors are pushing to release new software updates and additional functions desired by the end users, (2) end users are pushing to install software updates with minimal testing and downtime, and (3) systems integrators are being asked to provide a broader scope of services in less time.

Many, if not most, batch control systems are presently used to control processes in regulated industries. In these industries, the collection of data from the process; the reporting of operator actions, alarms, and events; and the capture of the sequence of operations from batch control applications is critical. Even in those industries that do not have to prove 100% data integrity, accurate data recording is important for optimization, efficiency calculations, and batch trend comparison.

The issues listed previously can lead to issues with data collection and data historians in many ways. Missed data due to application failure is common. Data that is "frozen" on a value due to transmitter errors or controllers that "hold last value" can skew data reporting and lead to questions from regulators. Every pharmaceutical, biotech, food and beverage, and medical device company can point to instances and cases where data were not collected due to one or more of these types of control system errors.

Most of these data recording errors occur because of data transaction and transmission problems, not because the data were not available at the source. "Holes" in historical data occur most frequently due to problems in the movement of the data values from the Level 1 sensors and data sources to the applications in the Level 2 or Level 3 application space that are charged with recording the data.

If the batch control system architecture in Figure 8.1 were redrawn to show the mapping of data transactions for the system, then it might yield the diagram shown in Figure 8.2. This figure shows how data would commonly be transacted between the various devices and applications in a batch control system.

Data transaction problems arise at "hand-off" points, locations in the data transmission route in which data are passed from one application to another. For example, an operator entry or temperature measurement is produced on the factory floor (e.g., transaction A in the lower center of Figure 8.2). The data are scaled or packaged by an automation controller and broadcast on the data bus or network (e.g., transaction B). The value is captured over the network by a SCADA application or DCS application that associates the data with a tag name entry, adds a time and date stamp, validates the data, and passes the data to the historian application (e.g., transactions C and D). The historian application, in turn, verifies the history group to which the data belongs, stores the value to a memory buffer, and periodically writes the data to disk (e.g., transaction E).

The data transactions described here are typical but could be much more complex in a given instance or application. Many additional transactions may occur between the source of the data and the final storage and use of those data. To complicate these transactions further, data may undergo transformation as protocol conversions are made in the transmission of data from the source to final use. Even in the simple case shown in Figure 8.2, issues often arise due to scaling, application errors, protocol errors, and disk subsystem errors, just to name a few.

Introduction to the DDTM

A new paradigm is emerging, one in which the "middleman" in these data transactions is eliminated, thereby eliminating the errors that occur as data pass through these intermediate applications, gateways, and protocol conversions. In this model, data are moved directly from the source to the final point of use, whether it is an operator console or a data historian archive on disk. Data do not pass through

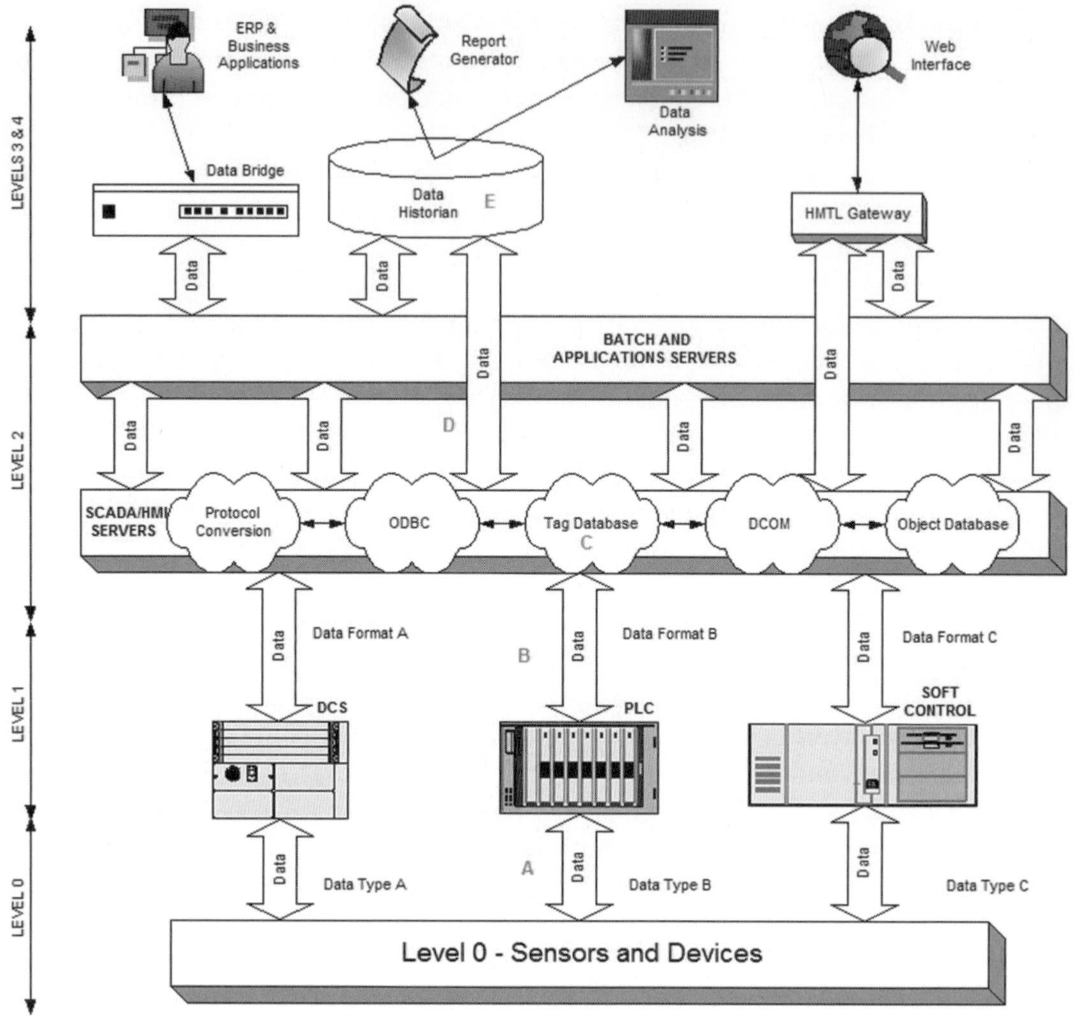

Figure 8.2. Common batch control data transaction mapping.

intermediate databases, protocol conversions, or intermediate applications. Data producers communicate directly with data consumers. A few of the new methods that are available include the following features:

- Use of a hardware interface at Level 1 devices that allow data to be moved directly into Level 3 applications
- Collection of real-time production data at Level 1 with subsequent batch transfer of data to a Level 3 data historian at a later time

- Use of embedded Level 2 and Level 3 applications in automation controllers that eliminate the need for network transactions for real-time data

- Use of a single application for process visualization, batch control, and manufacturing operations management, thereby eliminating the need for separate Human Machine Interface (HMI) and production management software

The DDTM methodology efficiently splits control applications between machine-level control requirements, cell-level control requirements, and enterprise control requirements. Unlike the current approach that makes data available to all control elements at all control levels, DDT forms requirements based on data producers and data consumers. The design of the control system, including hardware and application software, is structured to meet these specific requirements. This methodology is independent of any vendor's hardware or software offering. While the discussion here will focus on batch control and batch data, these methods can be employed on any process.

The application of this methodology was inspired by work done by various universities on the producer-consumer model in multiprocessing and sharing of resources in multiprocessor computer systems.[3] There is a parallel between efficient data production and consumption in process control systems and the efficient production and consumption of data in high-speed, multiprocessor computer systems[1].

In both process control and multiprocessor computer systems, data are consumed as "closely" to the source as possible, in order to minimize errors, outages, and issues related to application errors, transmission errors, protocol conversion errors, and hardware errors.

An Example of the DDTM

Using the example of a plant-floor temperature sensor producing data that ultimately is stored in a data historian, let's apply the DDTM to see how the architecture might be changed to improve data transmission efficiency and reliability. Assume that the temperature measurement is a monitor-only point (i.e., that it is not required for a control loop but is required for regulatory compliance). Also assume that it is a required element on a process graphic display and is required to be sampled and stored in history every 10 seconds. The process cell is controlled by a PLC and has an HMI and Batch supervisory system, as shown in Figure 8.2.

Before attempting to structure the solution, it will be useful to note that application of this model requires the design engineer to develop an orderly set of requirements. In order to do this, the engineer will need to understand the batch control system and tabulate all the producers and consumers of data. For our problem, the temperature sensor is the data producer, and there are three consumers, as shown in Table 8.1. Other signals have been added for purposes of illustration.

Remember, in the traditional control system architecture, this temperature measurement incurred five transactions to get from the plant floor to the data historian:

1. Sensor value is input into the PLC, and stored, scaled value is input into the I/O data table

2. Ladder logic, function blocks, and so on, are used to transfer data to the memory area used for block read to the HMI server

3. HMI server uses Object Linking and Embedding for Process Control (OPC) and others to read the scaled temperature from the PLC memory location and stores this value in a live tag name database (either locally or globally)

4. HMI server develops Open Database Connectivity (ODBC) packets to transfer the scaled temperature value to the Data Historian server every 5 seconds

5. Data Historian detects the presence of the data value and stores data in its internal database

Table 8.1. Producer and consumer table		
Item	*Producer*	*Consumer*
Temperature	Sensor–TE1004	PLC Input/Output (I/O) Subsystem HMI Graphic Display Data Historian
Flow	Mass Flow–FE1001	PLC I/O Subsystem HMI Graphic Display Data Historian PLC Control Loop Input
Flow Control	PLC Control Loop Output	FV1001
Flow Setpoint	Batch Control Server Recipe	PLC Control Loop Setpoint

Additional data transactions may be required for more highly distributed systems, systems that use software components from different vendors, or applications that require special formatting or special protocols.

With the DDTM, once the producers and consumers of data have been identified, the required architecture may be developed. Eventually, sensors and control elements will commonly have embedded intelligence that will allow them to initiate data transmissions. Until then, a PLC or DCS controller (or other intelligent I/O subsystem) will be required to read, scale, and make sensor inputs available for control loops and other consumers. Therefore, in our example, data transaction 1 will still be required, even though the temperature measurement is not required in the PLC.

The other two consumers of the temperature data are the HMI Graphic Display and the Data Historian. In the transactions listed previously, if the PLC to HMI server link is broken, neither the HMI subsystem nor the Data Historian will receive the temperature data. This is only one of five possible transaction failure points for this data. Only when all five transactions are "good" does the Data Historian receive its data. It was mentioned previously that in a regulated plant environment, a history of these data is critical (i.e., required for regulatory compliance). With five transactions between the point where the data are produced (i.e., TE1004) and the point where they are consumed (i.e., data historian), there is a significant risk of data loss.

One possible application of the DDTM using commercially available products yields the data transaction mapping shown in Figure 8.3, for the temperature example. Figure 8.4 shows the resultant network architecture after applying this model. In this example, the PLC was enhanced with an embedded data historian module and a database transaction module.

The embedded data historian has the capability to store data values at configurable time periods internally for several days. The data can be buffered locally and transferred periodically to the central data historian. An alternative configuration would be to transmit the data as polled with the local buffer used in case of transmission errors.

The database transaction module has the capability of reading data directly from the PLC memory, forming data transaction packets, and transmitting these packets directly to an HMI tag name database. This module can also be used to store operator actions and event data into the Batch server database.

Application of the DDTM is straightforward:

1. Identify the producers and consumers of data

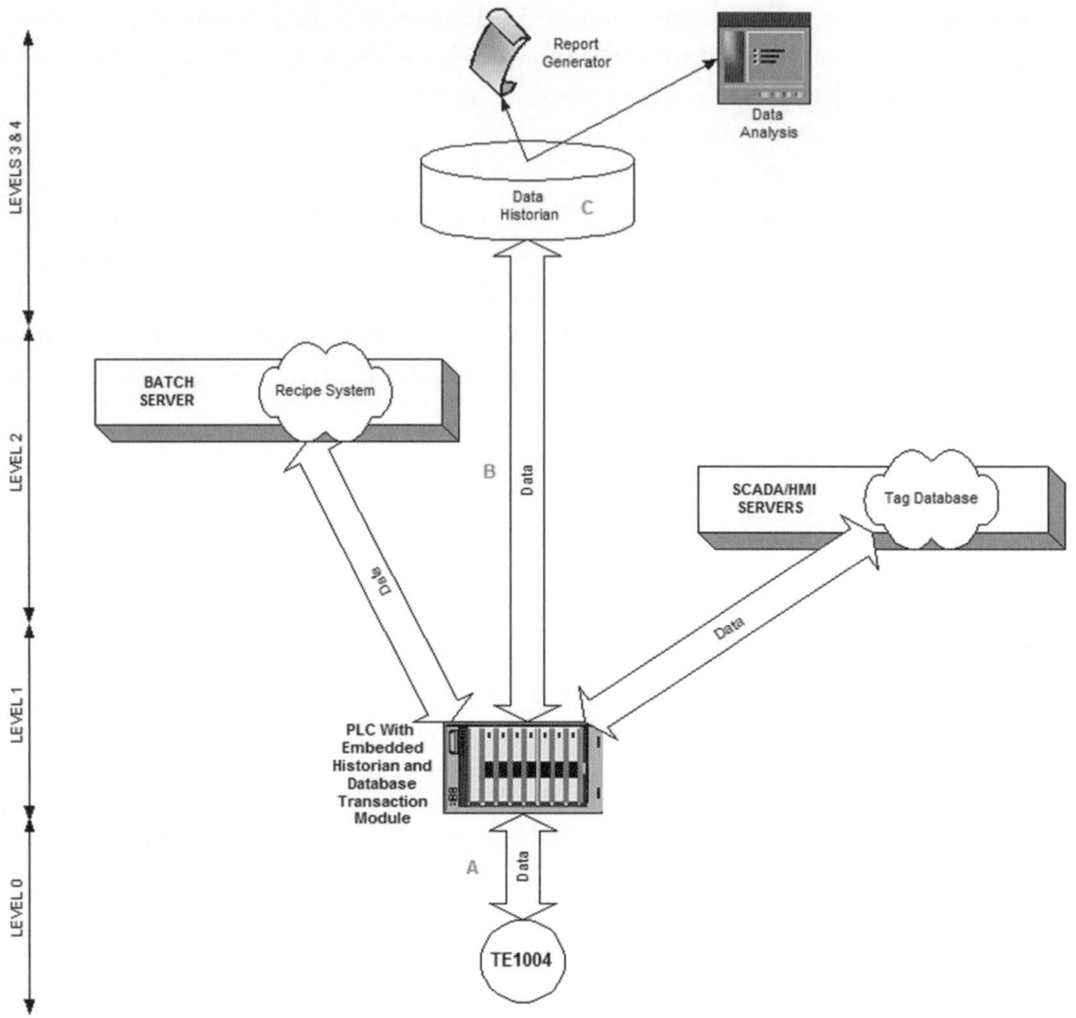

Figure 8.3. Batch control data transaction mapping after application of DDTM.

2. Select hardware components that will allow for direct data transactions between the producer and consumer

3. Construct the batch control system architecture from the selected hardware components so that all required data transaction paths are supported

As a result of applying this model, the number of data transactions in the temperature example was cut from five to three. This reduction in "hand-off" points can reduce risk of data loss and can increase overall process control system reliability.

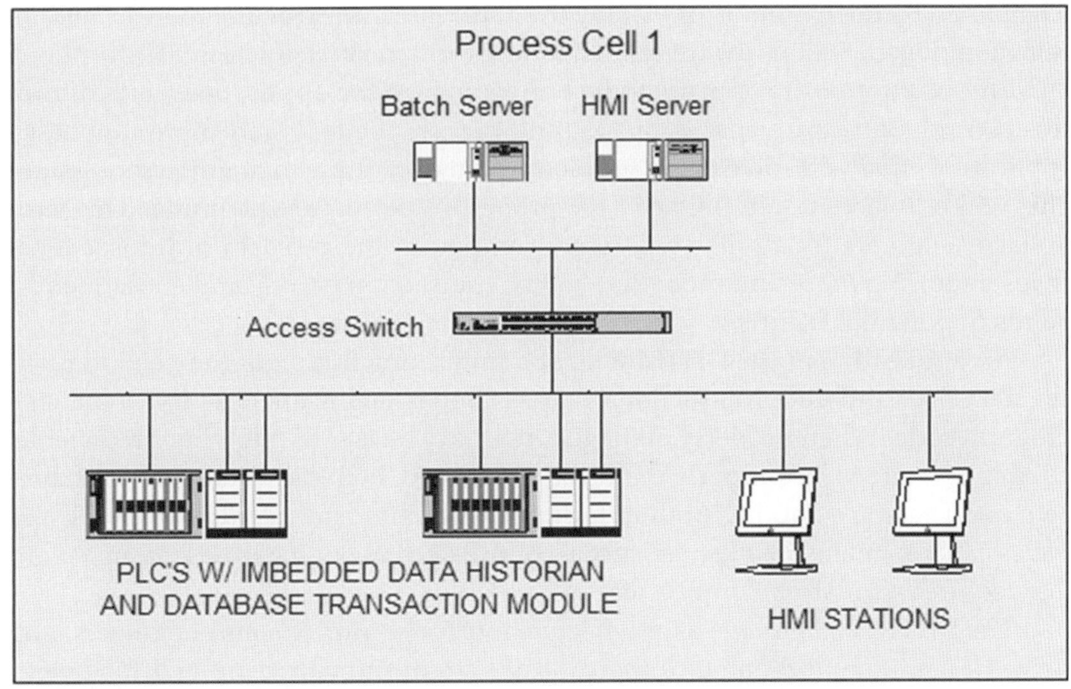

Figure 8.4. Batch control network architecture after application of DDTM.

Again, the first transaction (A) is sensing the plant floor variable in the I/O sub-system. This is an extremely reliable and robust transaction. The middle three transactions in Figure 8.2 (B, C, and D) were eliminated and were replaced by a single transaction (B, in Figure 8.3). In the case where data are buffered locally in the embedded data historian, this transaction becomes non-time critical.

The network architecture did not have to be modified significantly to accommodate the DDTM. Four additional Local Area Network (LAN) connections were made (Fig. 8.4) to accommodate the additional hardware modules added to the PLC subsystem. Additional efficiency and reliability can be achieved by smartly using redundant switches and virtual LANs. The DDTM yields an architecture that accommodates the definition of separate, dedicated virtual LANs for data historian transactions and HMI tag name server transactions.

The Future

More and more end users understand the need to install industrial-grade networking components in their plants, as opposed to office-grade components. Redundancy schemes and other methods for making the hardware infrastructure more robust are being employed, and hardware-related failures will most likely decrease over time.

The exact opposite appears to be happening on the applications side of process control. Software vendors are touting bigger, "better," and more numerous applications that will solve every problem, from downtime to maintenance issues. This will only accentuate the need for proper infrastructure planning. This rush to have "software fix all ills" will likely compound the problem of transactional chaos, requiring more and more data paths, transaction points, and consequently, points of potential failure.

An application of the DDTM will, at a minimum, highlight the need to manage data flow and data transactions, just as control engineers now try to manage program code. When properly managed, both can be assets, allowing predictable, reliable operations. Improperly managed or ignored, both can take on a life of their own and become a liability, both in terms of reliability and maintainability. These issues can also make life miserable for a control system engineering staff in a plant that requires regulatory compliance.

The DDTM will emerge as an effective tool for control systems engineers who strive to maintain and enhance the control system infrastructure and for operations personnel who strive to simplify and enhance their ability to interact with the equipment and process. Table 8.2 summarizes a few trends that I believe will continue in the process industries over the next five years. The DDTM supports each of these data-transaction minimizing trends. While only the passing of time will be able to validate this vision, it does make the future something to look forward to.

Conclusion

The DDTM described in this chapter is a methodology that yields a batch control network design to minimize the number of data transactions in the system while increasing reliability and data efficiency. This positive result is achieved by listing all data requirements by data producer and consumer, by selecting hardware components that allow for direct data transactions between the producers and consumers of data, and by constructing the network architecture to support these direct transactions.

This methodology is independent of the process type, instrument bus technology, control system hardware or software type, or communication protocol. With the wealth and diversity of data transaction, process control, batch control, and network hardware and software components on the market today, a solution to fit every application can be found by applying the DDTM.

Table 8.2. Possible process industry trends		
Present	*Trend*	*Future*
Central Data Historian collects data from multiple devices, controllers, and servers on Levels 1, 2, and 3	Move data collection closer to the sensors	Distributed Data Historians integrated with Level 1 controllers; use of Central Data Historians to collect data from distributed historian buffers on a periodic basis
Operations personnel interact with the batch control system through an HMI/SCADA subsystem; operations personnel interact with the business systems through a Manufacturing Execution System (MES) subsystem	Use a single-window interface for all functions	Single Operator Interface subsystem with one application to display production and process data in real time and also display and enter transactional business data
Sensor values are stored in a common HMI Tag Database and other applications are serviced (history, graphics, Statistical Process Control [SPC], MES, Enterprise Resource Planning [ERP], Web, etc.) from this database	Move data directly from source (producer) to destination (consumer)	Separate Data Transaction Modules at Level 1 controllers to move data to the appropriate Level 2, 3, and 4 databases and applications and vice versa.

References

1. Barr, Brian, and Priscilla McAndrews. 1996. Interprocessor communication. Interprocess Synchronization Tutorial, Center for the New Engineer, George Mason University.
2. Instrumentation, Systems, and Automation Society. 2005. *ANSI/ISA-95.00.03-2005: Enterprise control system integration, part 3: Activity models of manufacturing operations management*. Research Triangle Park, NC: ISA.
3. Verhappen, Ian, and Eric Byres. 2006. Industrial network integrity. *Intech* (October 2006).

CHAPTER 9

ISA-95 Implementation Best Practices Workflow Descriptions Using B2MML

Presented at the WBF
North American Conference,
March 5–8, 2006, by

Costantino Pipero
cos@beeond.net
Beeond, Inc.
1920 Fleet Street
Baltimore, MD 21231, USA

Kishen Manjunath
Solutions Architect
kishen.manjunath@honeywell.com
Honeywell Process Solutions
2500 West Union Hills Drive
Phoenix, AZ 85027, USA

Abstract

This chapter focuses on the best practices to describe workflows using Business To Manufacturing Markup Language (B2MML) applications based on the ISA-95 standards. ISA-95 and B2MML, like any standards, are subject to interpretation. Manufacturing Execution System (MES) and Enterprise Resource Planning (ERP) vendors could interpret these standards differently, which could result in variants of the implementations, thus leading to interoperability issues. This chapter outlines the best practices to describe workflows using B2MML applications that can aid in more uniform implementations, resulting in a reduced project implementation time.

B2MML applications illustrate how ISA-95 parts 1 and 2 can be applied to a usable Extensible Markup Language (XML) based infrastructure. These applications offer data mappings, description of the schema choices, how B2MML has been used, and how B2MML is evolving to support extensions and a broader range of transactions. Workflows illustrate how ISA-95 parts 3 and 5 are applied to describe interactions between B2MML applications and ERP systems in typical manufacturing operations activities. These implementation best practices are summarized with examples that include quality management, maintenance notification, production scheduling, and production performance reporting scenarios.

Introduction

What Does the Standard Provide?

ISA-95 parts 1 and 2 define the content and context of the information required for the interfaces between enterprise activities and control activities. ISA-95 part 3 shows activity models and data flows for manufacturing information that enable enterprise-control system integration. B2MML uses XML schema definitions to specify and constrain the content of the information flow between enterprise and control systems.

Why the Need to Define Implementation Best Practices?

The current definition of B2MML specifications is based on certain assumptions about enterprise activities and control activities. The dotted line in Figure 9.1 represents one way to separate MES systems below the line from ERP systems.

The interactions between these activities become more complex when considering the different production models:

- Make to stock

- Make to order

- Engineer to order

Agile production models require more transactions and more complex data to be exchanged between enterprise and control systems. Figure 9.2 illustrates this. Given these complexities and the current specifications of B2MML, there is a need to define implementation best practices that would enable a more uniform interpretation and use of the standards.

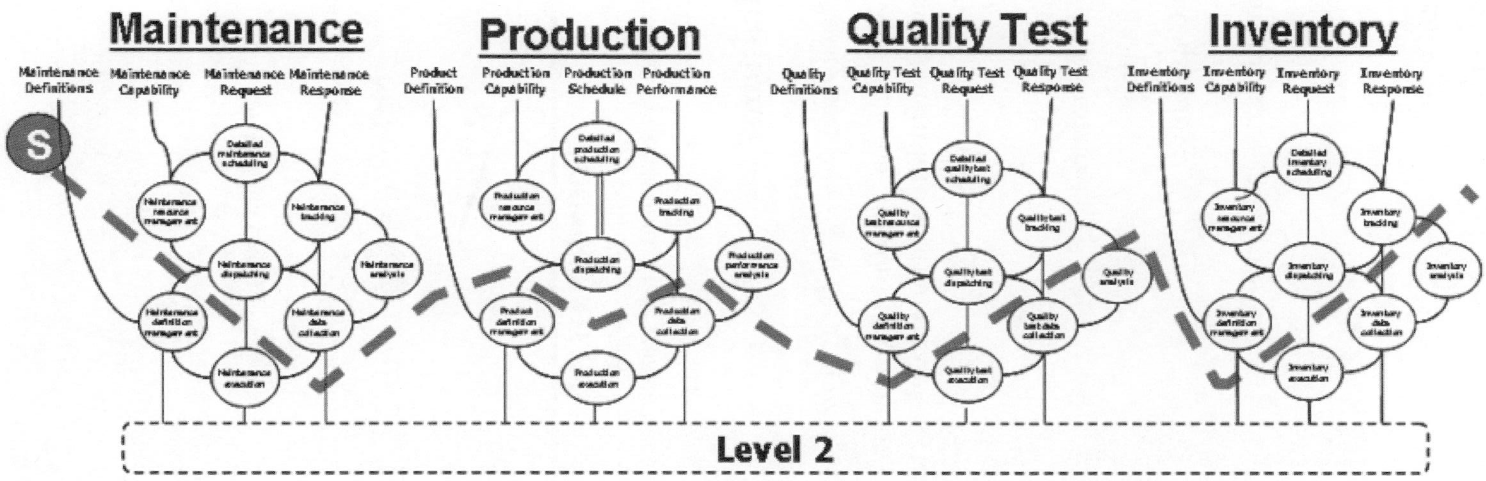

Figure 9.1. Activity model from ISA-95.03.

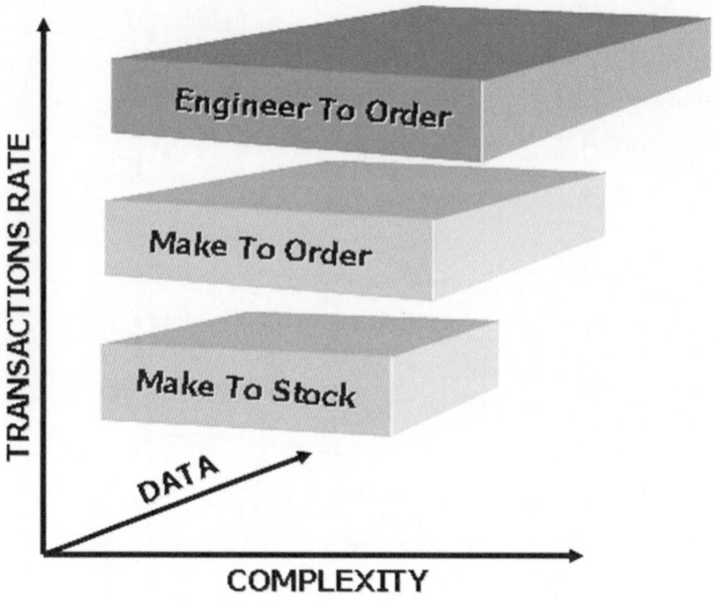

Figure 9.2. Agile production models require more transactions and data.

B2MML Applications

What Kind of Information Needs to Be Exchanged?

B2MML describes XML schemas for four information categories from the production operations management information models in figure 1 of ISA-95 part 2 and also shown in Chapter 16 (Fig. 16.1):

- Production Schedule
- Production Performance
- Product Definition
- Production Capability

It also describes four resource categories:

- Personnel
- Equipment
- Material and energy
- Process segments

A first approach is to determine which category of information needs to be exchanged in reference to the resources. How should one determine the context and content of the information that needs to be exchanged? It is important to understand the concept of the three distinct production models and the implications of these models to integration.

Make to Stock

This integration is probably the least complex in nature:

- Product definition and capability are usually planned before execution
- Product customization is limited
- Master data needed to perform production can be predetermined
- Master data need to be synchronized less often
    - Automatic realignment between planning and execution
    - Manual process
    - Communicated with the scheduling request

Production scheduling might happen over a large time window (e.g., weekly, biweekly) and can contain detailed scheduling information with multiple work orders to be dispatched to the shop floor. Performance can be reported at regular intervals, either at work order completion or at each execution step, depending on the procurement and replenishment strategy.

Make to Order

Make to order is often considered equivalent to a "make to stock" at zero inventories. Some features of this integration include the following:

- Product customization is important, and hence master data assume a more significant role. There is a more frequent flow of master data, which requires synchronization of the changes to product definition and capability.
- Product definitions are often contained within the scheduling requests. These requests are more granular with respect to time and less regular in frequency.

- Changes in the execution are usually notified after the completion of the operations or production performance results. In some cases, these changes may be notified earlier in the planning system, in order to resequence the plan and make the appropriate adjustments.

Engineer to Order

This is probably the most demanding integration:

- Information exchange is synchronous with minimum delays.
- The aerospace industry offers one of the best examples. Products are designed and engineered to drive the planning, but these definitions could change during execution.
- Change requests might be sent back to engineering or planning, which puts the execution on hold, waiting for the changes to be reflected in planning. A new set of operations or a variation on the route (e.g., in case of added inspections) will then be communicated to the shop floor to complete the work orders.
- Messages are richer in content and require a significant amount of extensions to B2MML.
- Performance reports contain as-designed and as-built data.

How Is the Context Aggregated in Such Complex Structures?

A key concept to understand is the use of segments in the B2MML schemas. A segment is a functional unit of integration and defines the content model for information interchange.

The segment requirement element encompasses the aggregation of all the production- and process-related information within one schema to be transacted through activities such as production schedule and production performance.

Typically, a segment maps to functionalities like operations, unit procedures, and phases of execution; it reports information like equipment and resource allocations and material inputs and outputs.

Correlations among Different Schemas

Some examples that show the correlations among the different schemas and how they represent different contents of the same information are as follows.

The user can enter specific structural information in the product definition schema. For example, Figure 9.3 shows how the material specification information is organized within the product definition.

The following XML snippet is a possible interpretation of the structure where a class of materials, a specific inventory lot of that material, and a Work-In-Progress (WIP) sub-lot of the item are described:

```
<MaterialClass>
    <ID>Sug001</ID>
    <Description>Pure Cane Sugar Type 1</Description>
    <MaterialClassProperty>
        [ . . . ]
    </MaterialClassProperty>
    <MaterialDefinitionID>String</MaterialDefinitionID>
    <Any/>
</MaterialClass>
<MaterialDefinition>
    [ . . . ]
    <MaterialClassID>Sug001</MaterialClassID>
    <MaterialLotID>Lot_Sug_123</MaterialLotID>
    <Any/>
</MaterialDefinition>
<MaterialLot>
    <ID>Lot_Sug_123</ID>
    <Description>Domino Cane Sugar</Description>
    <MaterialDefinitionID> . . . </MaterialDefinitionID>
    <Status>Ready</Status>
    <MaterialLotProperty>
        <ID>PlannedQty</ID>
        <Description>Expected Quantity at hand</Description>
        <Value>
            <ValueString>200</ValueString>
            <DataType OtherValue="String"
EnumerationID="String">dec</DataType>
            <UnitOfMeasure>KG</UnitOfMeasure>
            <Key/>
            <Any/>
        </Value>
        [ . . . ]
    </MaterialLotProperty>
    <MaterialSubLot>
        <ID>WIP_Sug_001</ID>
        <Description/>
        <Status>Ready</Status>
```

Figure 9.3. A different schema. (*continued on next page*)

```
        <StorageLocation>Silo1</StorageLocation>
        <Quantity>
            <QuantityString>20</QuantityString>
            <DataType OtherValue="String"
EnumerationID="String">string</DataType>
            <UnitOfMeasure>KG</UnitOfMeasure>
            <Key/>
            <Any/>
        </Quantity>
        <MaterialLotID>Lot_Sug_123</MaterialLotID>
        <Any/>
    </MaterialSubLot>
    [ . . . ]
```

Figure 9.3. A different schema. (*continued*)

The system will also have to know specific capability information about that material (e.g., its availability and use). Figure 9.4 shows where this information can apply:

A possible XML instance might be as follows:

```
<MaterialCapability>
  <MaterialClassID>Sug001</MaterialClassID>
  <Description>Pure Cane Sugar</Description>
  <CapabilityType>Committed</CapabilityType>
  <Reason>String</Reason>
  <EquipmentElementLevel>Area</EquipmentElementLevel>
  <MaterialUse>Consumable</MaterialUse>
  <StartTime>2001-12-17T09:30:47-05:00</StartTime>
  <EndTime>2001-12-21T09:30:47-05:00</EndTime>
  <Location>
    <EquipmentID>Blenders</EquipmentID>
    <EquipmentElementLevel>Area</EquipmentElementLevel>
  </Location>
  <Quantity>
    <QuantityString>200</QuantityString>
    <DataType>dec</DataType>
    <UnitOfMeasure>KG</UnitOfMeasure>
    <Key> </Key>
  </Quantity>
  [..]
</MaterialCapability>
```

Figure 9.4. A material schema.

It all comes together in the production schedule, where the master data are referenced in a production request. Figure 9.5 captures an excerpt of the segment requirements, where a material has been planned and released to production.

Workflow Descriptions

Why Are Workflows Complex?

As per the ISA-95 part 1 functional data flow model, there are nineteen different data flows of interest between the enterprise and control system:

- Schedule
- Production from plan

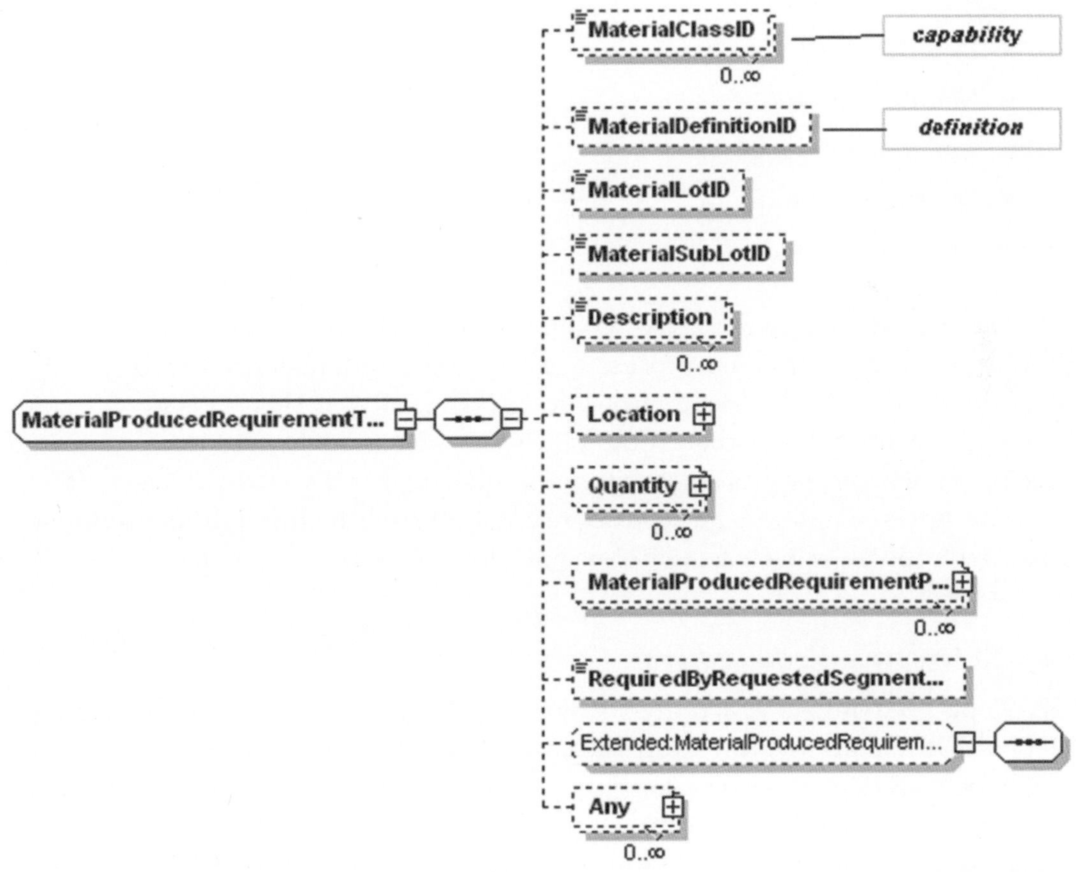

Figure 9.5. A material requirement schema.

- Production capability

- Incoming order confirmation

- Long-term and short-term material and energy requirements

- Material and energy inventory

- Production cost objectives

- Production performance and costs

- Quality assurance results

- Standards and customer requirements

- In-process waiver request

- Finished goods inventory

- Process data

- Pack out schedule

- Product and process know how

- Maintenance requests

- Maintenance responses

- Maintenance standards and methods

- Maintenance technical feedback

There are three different production types (e.g., discrete manufacturing, batch processing, and continuous processing) and three different production models (e.g., engineer to order, make to order, and make to stock). This leads to nine combinations (or hybrids) of the production types and models, with each having a specific set of business processes within the enterprise and control domains. When the data flows of interest are considered within each of the hybrid production types and models, it leads to complex workflow descriptions.

How to Describe Required Workflows?

ISA-95 part 3 defines the activity models of Manufacturing Operations Management (MOM) that enable enterprise and control system integration, as shown in Chapter 16 (Fig. 16.1). The key to defining the information aspects of manufacturing operations is capturing the context and content of data that need to be exchanged—this is what gets defined as B2MML applications. B2MML applications, in conjunction with the transaction models defined in ISA-95 part 5, provide a basis for describing the workflows. Some examples of workflows for the different production types are illustrated in Figures 9.6, 9.7, and 9.8.

Conclusion

The key to defining B2MML applications is identifying the context and content of the information that needs to be exchanged. Workflow descriptions using B2MML vary in complexity, depending on the production model, and these workflows require identifying the B2MML elements that would be used to correlate data that flows between enterprise and control systems. B2MML applications, in conjunction with the transaction models, provide a starting point for describing the workflows in enterprise-control system integration.

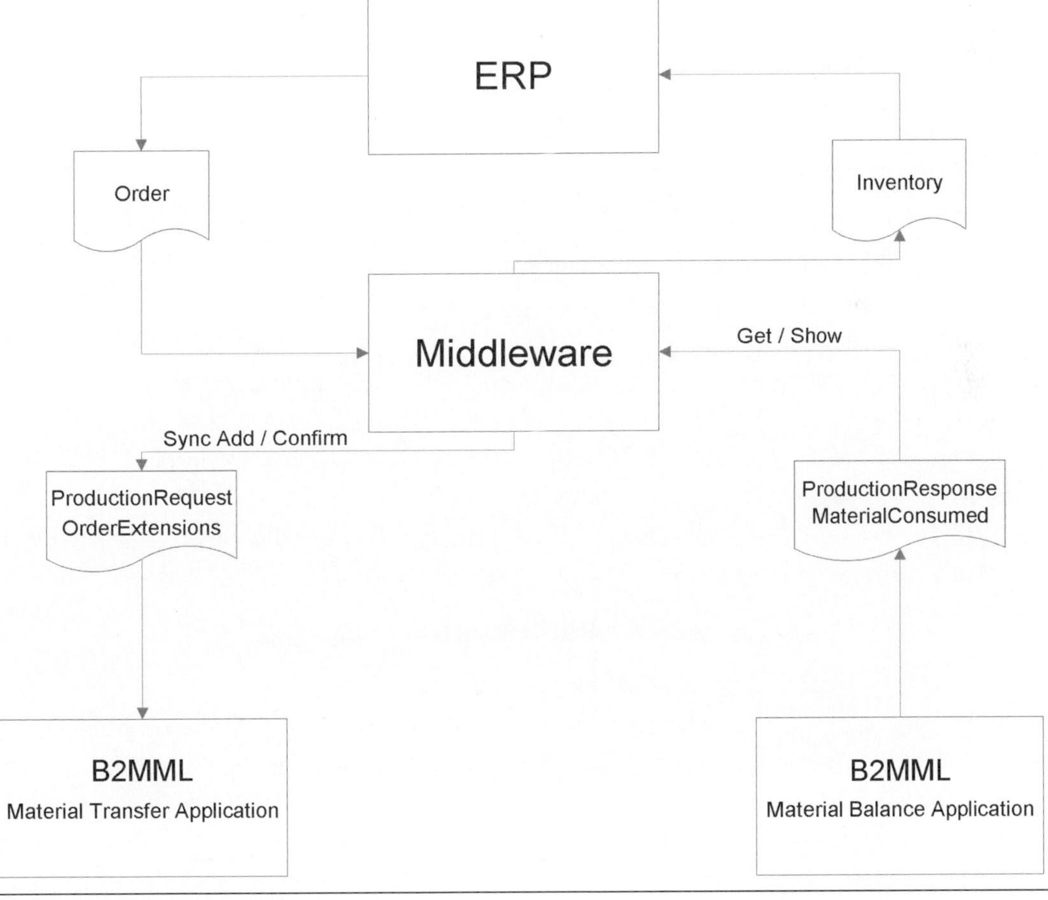

Figure 9.6. Continuous process workflow.

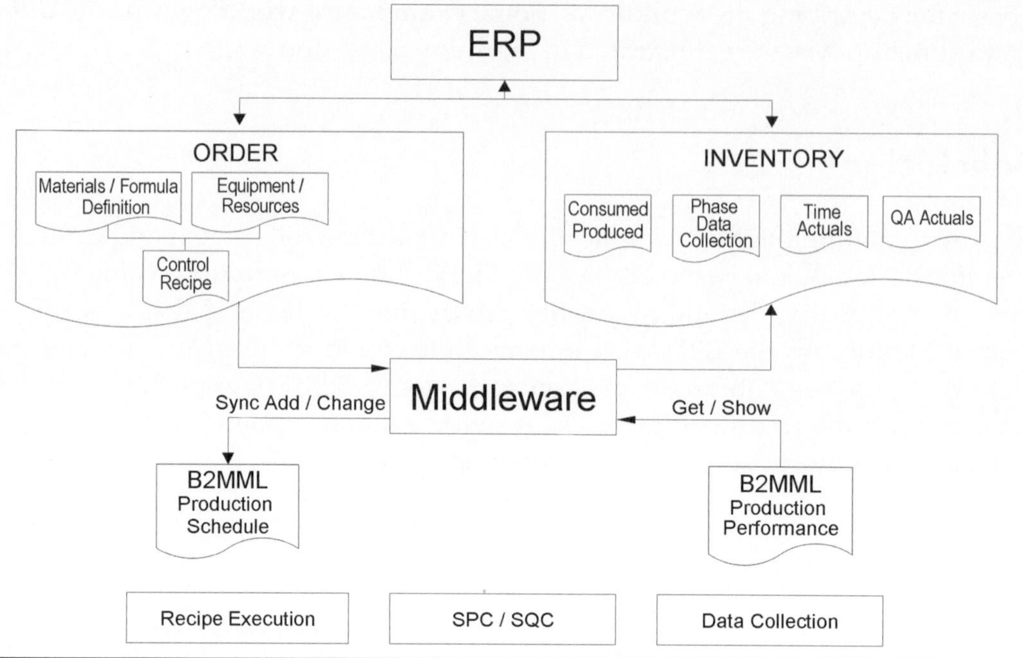

Figure 9.7. Batch processing workflow.

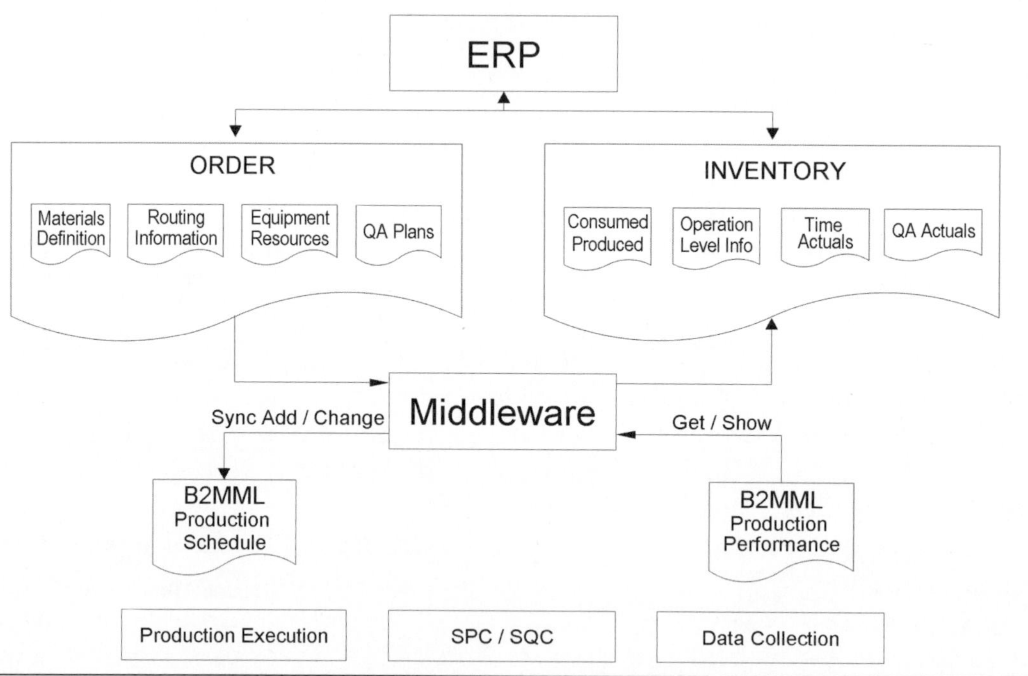

Figure 9.8. Discrete manufacturing workflow.

Translating Process Control System Data into Business Decision Information in Manufacturing Facilities

Presented at the WBF
North American Conference,
April 30–May 3, 2007, by

Ron Huffman
Director of Operations Analysis
ron.huffman@vigilistics.com
Vigilistics, Inc.
23361 Madero Street
Suite 100
Mission Viejo, CA 92691, USA

Abstract

Manufacturing industries have been reviewing, validating, and implementing various levels of process control automation for several decades. While this implementation of technology is credited with improving the overall control of operational processes, there continues to be a gap between the data generated by the technology that controls manufacturing processes and the information required by the business-level decision makers. While the amount of data generated and collected during the manufacturing process is greater now than it has ever been before, the amount of accurate manufacturing information has not kept the same pace with what is required for timely business-level decisions. Using currently available technology, the effort required to modify control systems so that the data can be collected and correlated with actual operational events can be extensive. In many cases, the risk of disrupting operations due to the effort required to modify the current control systems has caused many manufacturing operations managers to continue with their current manual data collection and reporting methods.

A much lower-risk packaged technology has been developed that layers over existing plant automation and provides the benefits of an ISA-88 based product record without the expense and risk associated with reengineering plant automation systems. This technology uses a data collection model in addition to an ISA-88.01 physical model to collect all phase-event information and process-generated data in a full-context and precorrelated format. All data are recorded in an open SQL database and are available for use immediately upon collection. The specifics of the technology, barriers that can be overcome, system benefits, and examples of the business information provided are all described in this chapter.

Plant Control Technology

Most manufacturing industries have been reviewing, validating, and implementing various levels of process control automation for several decades. In most cases, manufacturing plants have implemented various levels of the following technologies in order to improve control of manufacturing processes:

- Sophisticated process control systems to automate processes and reduce the amount of human intervention required to properly operate a process

- Human Machine Interface (HMI) screens to provide operators with the ability to monitor and control the various processes to the best capabilities that a human can provide

- Batch Control Systems to automate recipe handling, automate the control of batch processes, and record batch process results

- Data historians that record the control system point data generated during the various manufacturing processes

All this implementation of technology is credited with improving the overall control of manufacturing processes. However, there continues to be a gap between the data generated by the technologies that control manufacturing processes and the information required by the business-level decision makers to optimize the business profitability.

Business System (ERP) Implementations

While plant control system enhancements were taking place, business process systems, primarily with an Enterprise Resource Planning (ERP) focus, have been implemented. The intent of ERP is to integrate all the business information systems of any company (e.g., Financial Reporting, Procurement, Accounts Payable, Accounts Receiveable, Manufacturing). ERP systems also serve to increase the timeliness and accuracy of the business information by eliminating "islands" of information where multiple systems have been implemented over time to perform specific business function tasks (e.g., one system for collecting and recording employee data, another system for performing accounts payable functions, yet another system for performing corporate purchasing functions, etc.).

These different business systems have two things in common: they record data in a transactional format, and they keep the data in a transactional database. Business systems typically require the final results of any process (e.g., total units produced, total quantities of materials utilized, total time to produce, etc.). For example, the total units produced for each batch created is needed to calculate the batch yield as well as to calculate the raw material usage for inventory control purposes. This information is required by the business system to generate the corresponding raw material usage and inventory reconciliation reports.

Each business process database record includes the relevant information concerning the completion of the particular type of transaction. For example, a receipt of raw materials will typically include the information concerning the transaction made between the receiving company purchasing agent and the supplier that provides the raw material. These would include the supplier name, raw material ID, order quantity, price per unit, delivery data, payment terms, and so on. These transactional data are based on accounting requirements to monitor the business and provide information about how to best operate the business. Once the data has been recorded in the business system database, it is available for reporting and analysis purposes virtually forever.

ERP systems provide the advantage of single-source information system architecture for collecting these transactional data, as opposed to having similar transactional data that reside in completely different systems that may not communicate with each other.

However, since most business systems provide information based on the completion of accounting transactions, and most plant control systems and plant data collection systems provide data that is time based and not transactional in nature, most ERP system implementations have not included tight integration with the manufacturing process control systems. Thus there is a gap between manufacturing (i.e., plant floor) data and business systems.

Manual Data Collection

To compensate for this information gap, most manufacturing operations typically create manual records of the operational information as it occurs during the process (e.g., start production time, end production time, clean-in-place time). All relative quantitative data generated during the process (e.g., final temperature, tank level, volume of product produced) are usually manually recorded by operations personnel in order to satisfy the requirements of the particular business system needs. The manually collected data are usually keyed into the business system later.

However, a manual data-collection process cannot possibly collect and correlate all actual events and all quantitative data point values as accurately as an electronic system. There is just too much data generated at too high of a frequency for humans to collect consistently or accurately.

The lack of electronically collected and recorded actual operational process event information by the control system is ironic. Where better to generate a record of the actual operational events as they occur than by the system that monitors and controls the operational events in the first place? The control system knows exactly when it must change operational states and what the measured process values are at any point in time. Control systems must know this to operate and control a process properly.

While control systems are typically programmed to control process changes in the most logical fashion and at a very high rate of speed (e.g., at a millisecond level), control systems typically do not include the data-collection logic required to record the actual operational events that a particular piece of equipment is transitioning through at any given time (e.g., startup, shutdown, production, idle). Control systems do an excellent job of consistently monitoring and controlling processes, but they typically are not programmed to "remember" what they just did or why they just did it at any point in time.

Analysis of Available Historical Data

The installation of data historians has helped make quantitative, historical manufacturing data available for analysis purposes. However, in most cases, the data collected by data historians are usually individual quantitative point data (e.g., temperature, pressure, tank levels), collected with a timestamp associated to each point collected. The events occurring in the process at the exact same time the point data were collected are typically not included. Even if the events were recorded, there usually is no mechanism to automatically correlate the operational event activity with the point data. While having the individual point data available in the plant data historian makes it relatively easy to generate very nice graphs and trend charts of the individual data points, not knowing what was actually going on in the operational process when the data were being collected makes interpreting the data difficult.

For example, a batch of high-value and high-cost product has been determined to be out of tolerance for acceptance by the customer, and the events that led to the loss of the batch need to be investigated. A data historian may do an excellent job of collecting all the electronically created temperature and pressure data associated with the batch processing vessel. Very nice graphical trends of the various data points may give insight into what caused the problem. However, knowing how the batch process was operating at the exact same time that any given data point was being collected provides much more useful information for an analyst. The point data may give the analyst one view of what caused the problem, but also knowing the actual events that occurred during the creation of the batch, as provided by the control system, will give the investigation team a very clear understanding of what actually caused the batch to go bad. The ability to review the same correlated data for all upstream (e.g., raw material data, pre-batch processing) and downstream processes (e.g., destination storage vessel condition, packaging process) that were directly related to the bad batch provides an even clearer view of what could have caused the batch to go bad.

While operations collect most, if not all, the critical data associated with each batch produced, the data may be collected using multiple methods (e.g., manually collected batch sheets, Microsoft Excel reports, automatically generated trend charts), and all the data may not be available immediately.

Manually collecting information for each batch produced using a "standard" paper document, where the batch process operator writes down the required information collected from a variety of sources (e.g., pressure and temperature gauge readings, HMI screen displays, the clock on the wall) and initials the form, is common. However, the accuracy of manually collected data is usually not worth the scrap paper it's printed on. Having more than one source of any electronically collected data, such as a batch control system for recipe management and another electronic record system to collect the actual ingredient quantities, is common as well. The correlation of these data from many sources can be difficult and time consuming.

If it is determined that the cause of the bad batch was an ingredient, then the need for the actual information (e.g., ingredient storage tank identification, lot numbers, Quality Assurance [QA] sample results) of each ingredient needs to be located and included in the analysis. If there is a possibility that the tank from which the bad batch sourced the ingredient can have more than one lot of the ingredient at the same time, then the complexity of the analysis increases. In the case of a bad batch, there are many questions that need to be answered, depending on the operation and the consequences of the batch. For our bad batch example, some potential questions are as follows:

- In the case of temperature-sensitive bulk ingredients, what was the temperature of the ingredient at the time that the ingredient was loaded into the batch tank?
- Did the temperature of the ingredient in the source tank of the batch ever exceed acceptable limits prior to being used in the bad batch?
- What was the condition of the batch tank immediately prior to starting the batch process?
- When was the batch tank last cleaned? (When was the cleaning previous to the batch?)
- Was the batch tank properly cleaned? (Was the previous cleaning completed as planned?)
- If the problem with the batch requires a recall, then what were the immediate destinations (e.g., storage tanks, packaging equipment) of the batch once it left the batch tank?
- Is it possible that there are there any remnants of the bad batch left in any of the destinations?

Unfortunately, using some of the existing technologies, it can be extremely difficult to get all the required information to answer these questions from a single data source for the individual batch. Correlating what was going on upstream of the batch, downstream of the batch, or in a related process at the same time when the problem occurred is more difficult. This is especially true when the upstream, downstream, and parallel process information are not recorded at all, since that information may not be critical for business system reporting.

Usually, it is up to the person(s) performing the analysis to determine the events based on their knowledge of the process or by reviewing manually collected data concerning the actual operational events, in addition to the data provided by the data historian. This is especially difficult when the data analysis is being performed at a different physical location or when the analysis is comparing the data from more than one facility at the same time or if the quality of the manually collected event data is suspect in the first place.

Collecting manufacturing data utilizing a combination of manually collected data and electronically generated data definitely does provide insight into operation activities. This method does satisfy the data requirements of business reporting systems. However, being able to have correlated operational event information generated by the systems that control the process in combination with the process measurement data collected by the control system gives the greatest insight into what actually happened at any time during the process.

To achieve this level of data collection and correlation, knowledge of the actual process must be included in the system that performs that data collection process. In order to best understand a manufacturing process and all the processes around it, a model of the physical processes in relation to the data generated by the processes must be created.

It All Starts with the Model

Until the early to mid-nineties, a standard methodology of modeling the operational states for manufacturing processes did not exist. The ISA-88 Batch Control standard, published in 1995, provides a methodology for designing and operating batch processes. However, ISA-88 is flexible enough to work quite well with discrete and continuous processes. Basically, ISA-88 provides a methodology to think through, and accurately model, the operational states that each piece of the process transitions though and how to define what triggers the transitions from one operational state to the next.

Adding a layer of knowledge to the model that includes all possible flows of product from each operational unit to another, as well as the types of flow possible (e.g., one to one, many to one, one to many, many to many) based on any process constraint for each unit (e.g., product type, upstream operational phase, downstream operational phase), provides a detailed map of how product can flow throughout the process and how the processes are related.

The addition of a layer describing the data to be collected for each process unit and during each process operational phase completes the model. Once completed, the model provides a graphical description of each process function, including the operational units, the operational states for each unit, the possible flows in and out of each unit, the possible upstream and downstream units related to each unit, and the data being collected for each unit (Fig 10.1).

Clearly defining the operational states for all processes is very important. For all manufacturing processes, properly determining when any unit is in a *production* state (i.e., actually producing product) versus all *non-production* states is critical to properly reporting what is actually happening in the manufacturing process.

Being able to collect and correlate measured consumption (e.g., energy, raw materials) in production states of operation versus non-production states of operation gives a very clear picture for calculating true asset utilization and actual energy consumption when in production and non-production operational states.

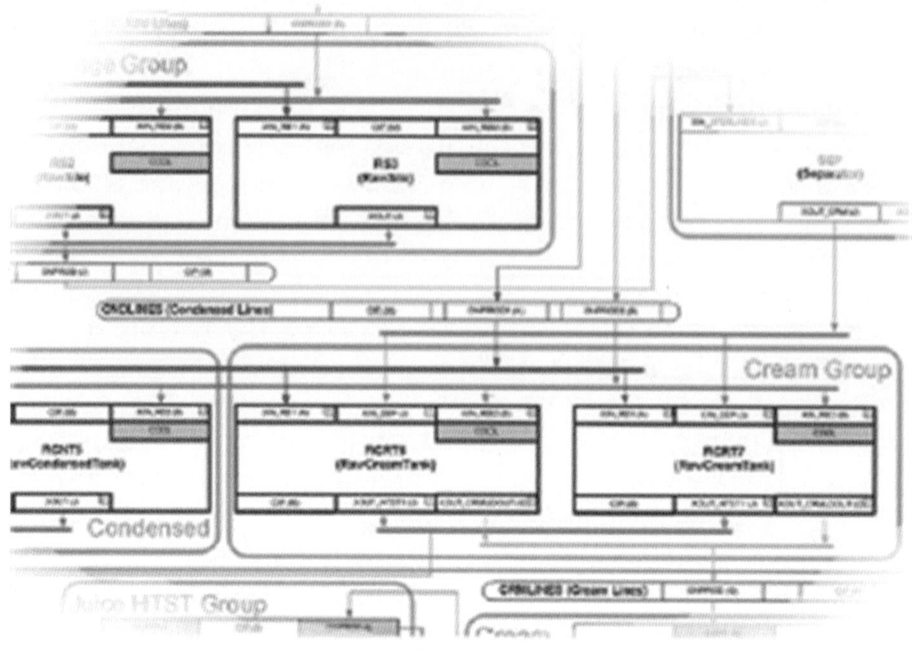

Figure 10.1. Process model example.

Defining States (Production versus Non-production)

The detail to which the Non-Production states can be modeled is up to the both the expertise of the people asked to provide the details of each operational state diagram and the extent to which the combination of data sources can provide the mechanisms to consistently move from one state to the next.

For example, a finished-good packaging operation wants to add an operational state for downtime. It is determined that each packaging unit can have the following causes for downtime.

- Out of packaging material
- Operator-initiated stop
- Power failure
- Unknown mechanical failure

Each individual cause of the downtime event can be determined based on information available in the control system. If the model includes each of these as individual downtime states, the system will automatically be able to collect and report exactly how much time each packaging machine spends in each operational state, over any range of time.

This ability to model the operational states to the greatest beneficial level of detail, in combination with modeling the data being collected during each operational state, provides a much clearer picture of what can be occurring during any manufacturing process, or collection of processes, at any given time.

Once the physical flow model is complete and validated, the information provided is used to configure the software that will connect to the various data sources and collect all the data as modeled. The potential sources of the data can include control systems, manually entered data, other databases, or any combination at the same time. In fact, the software acts as a "universal translator" for the data generated and collected from any and all data sources (Fig. 10.2).

In the case where the control system is a source of the data, small software agents are created and integrated into the existing control system without any change to the actual process control side of the control system. These agents have no effect on the actual control of the process. Rather, they just collect the actual state transition times to the millisecond for each process unit and collect the actual measured data associated to each individual process (e.g., temperatures, pressures, flow rates, energy consumption).

All control system data are collected in real time, using a single clock in order to record the process data and event correlation as accurately as possible. Now

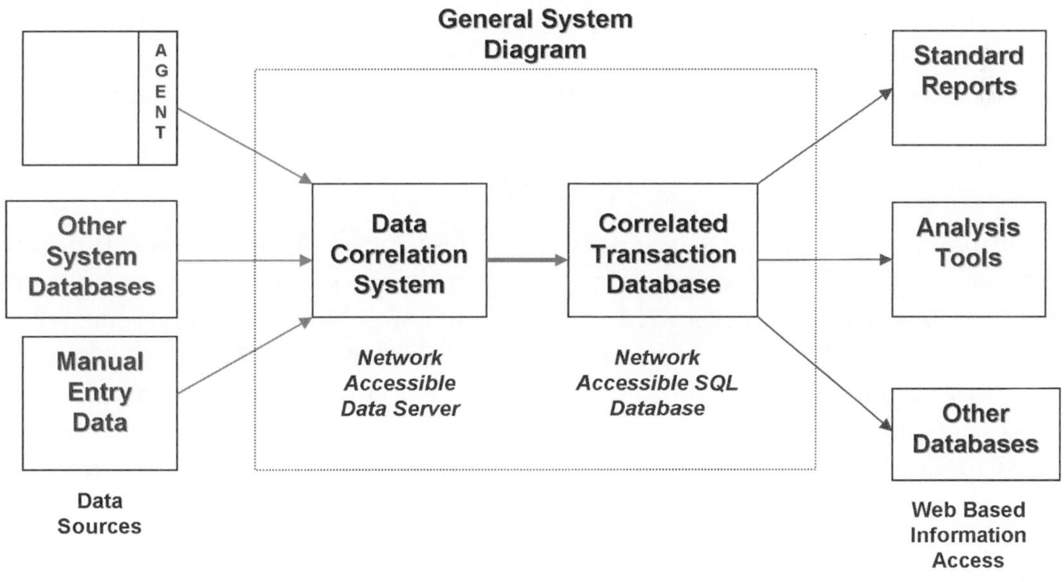

Figure 10.2. General system diagram.

that the manufacturing data are collected and correlated to a single clock, it can be recorded into a standard SQL database in real time. Once the correlated data are in an SQL database, they can be utilized using queries, just like any other transactional database data.

Minimizing Risks of Implementation

One of the biggest barriers to implementing this robust type of data collection capability is the risk of disrupting plant operations, due to the changes that many technology vendors require for existing plant process control systems. While manufacturing plants have varying levels of satisfaction with the current functionality of their control systems, most would be hesitant to risk the possibility of plant shutdown or substantial decrease in productivity during the implementation period of the robust data collection and correlation capabilities described previously.

A large benefit of the technology described is the minimal risk of disrupting current control system functionality, while still gaining all the benefits associated with correlating operational events and accurate process data into a transactional data format.

Benefits to Operations

Just as ERP systems provide the advantage of a single source of correlated transactional data for all business systems, having actual manufacturing data recorded in a transactional database format, based on the actual process model and flow paths, provides the following advantages (as described in the sections that follow).

Real-Time Actual Operations Reporting

Typical operations reports provide a written description of the events and timing of events. This manual description includes human interpretation of the events and results. Manual data collection cannot happen in real time, and manual reports can only be written as soon as the opportunity to write them is available.

This system provides the most timely and accurate information using the control system's perspective of the events. In this system, the data collection and correlation process happens in real time, and the production reports are immediately available as soon as the events occur, over any range of time.

Actual measured results, correlated with the actual process operational states, provide for much more timely and accurate reporting of manufacturing process information (Fig 10.3). Some examples include the following:

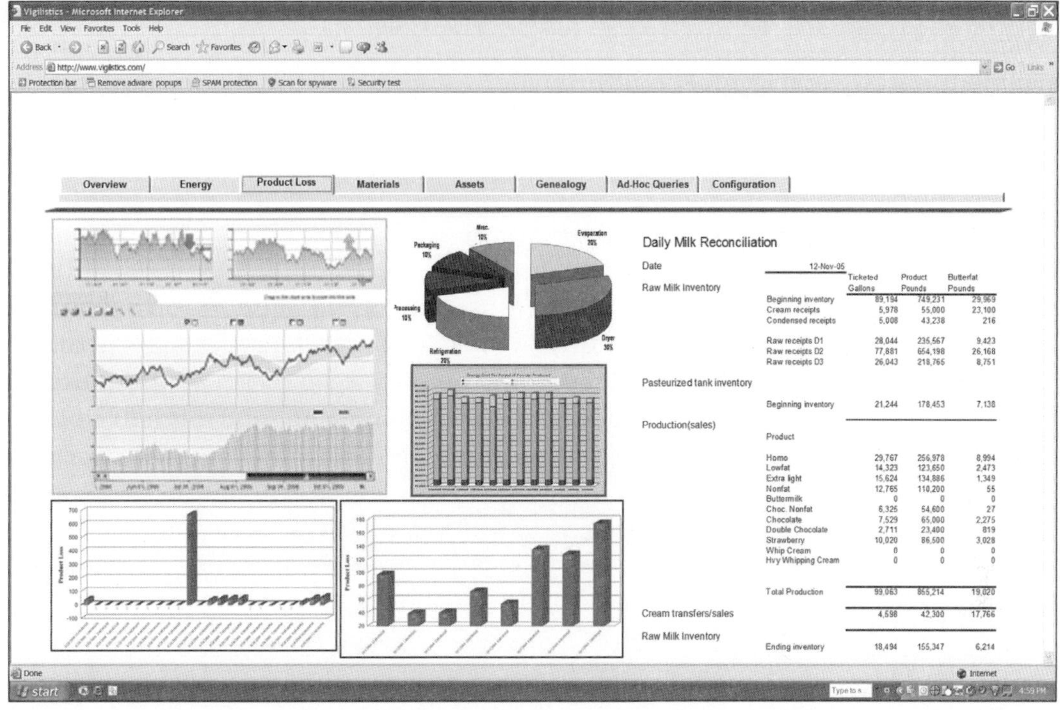

Figure 10.3. Operational report screen examples.

- Real-time operations results for any range of time (e.g., hour, day, week, shift)

- Actual asset utilization reports in real-time

- Actual comparison of activities in Production versus Non-Production modes of operation

Lot Genealogy

Since all possible product flows are known by the system from the process model, and since the actual flows are recorded as they occur, the system quickly provides the actual lot genealogy of any finished product, through any piece of equipment at any time. This also allows accurate tracing of potential contamination of other products.

The data collected, based on the plant model and actual flow data, allow the user to trace products forward and backward through the plant storage units, processing equipment, valves, and pipes. The graphic depiction of genealogy is easy to understand and navigate (Fig. 10.4).

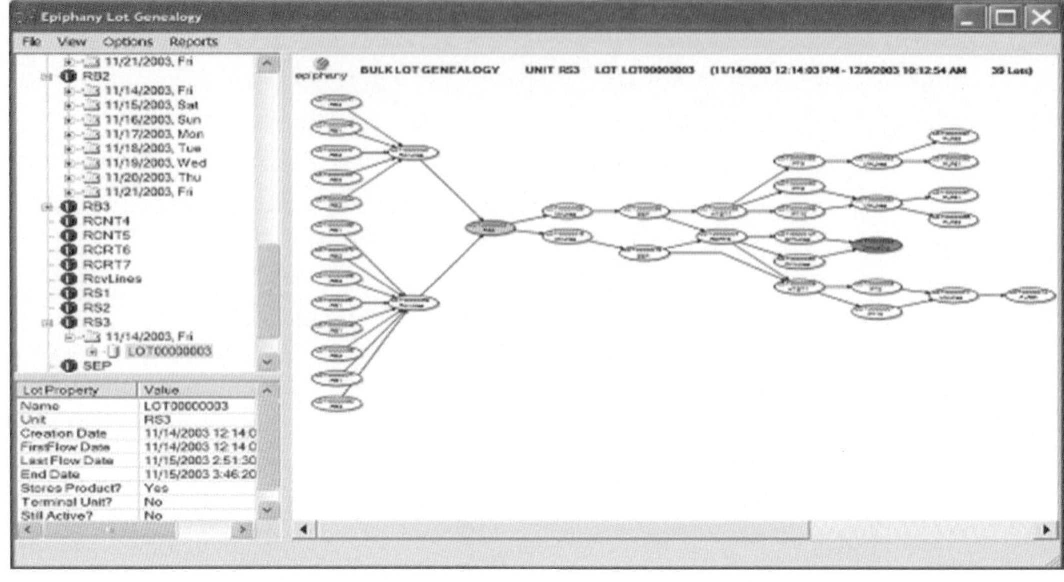

Figure 10.4. Lot genealogy example.

The user of this interface can trace any material or ingredient forward and backward from any point in the process model. The correlation of data allows the user to search for any attribute of a material that has entered the system. The system is configured to collect lot traceable parameters such as barcodes, bill of lading numbers, or operator names. These parameters can be searched and instantly a graphic visualization appears that shows the point of origin of the searched parameter. All product lots that could have been affected by contamination are shown in the diagram, as well as the path through the plant taken by the parameter. All commingling, by-products, and mixed lots are shown as well.

Adding the ability to know when a piece of equipment is washed or sterilized gives the system the ability to display all possibilities of contamination after a known contaminant passed through any modeled piece of equipment. This inherent knowledge of possible and actual flows makes determining upstream and downstream transitions and all possible contaminations much easier than using traditional manual tracking and tracing techniques.

Real-Time Product Loss Calculation

Being able to understand the actual paths of flow throughout the various processes provides the basis for lot genealogy, as described previously. Adding electronic measurement data of product flows through the process increases the ability to mass balance any section of the operation where the inputs and outputs are measured.

Since the system is collecting and recording all measured flows in and out of each piece of equipment using the same system clock, it is able to determine the total flows into a process unit, as well as all flows out of a process unit, with the difference being the product loss or gain for any slice of time (e.g., every minute, hour, day) or series of time slices (e.g., each hour for the past 12 hours). In the case of a storage vessel (e.g., a product storage tank), if a level sensor is attached to the storage vessel, the change in level can be included in the mass balance equation to more accurately determine loss or gain (Fig. 10.5).

Integration with Business Systems

Since the actual manufacturing results are recorded in a transactional database format, SQL queries can be developed to "boil down" the large quantities of control system data so that only the information required by the business systems can be provided by this system.

For example, this system will record and add up all periodic finished product flows through the manufacturing process. The final quantities produced will be recorded immediately upon completion of the production phase of any production unit. The final result of a process is typically what is required in a business system for reporting purposes.

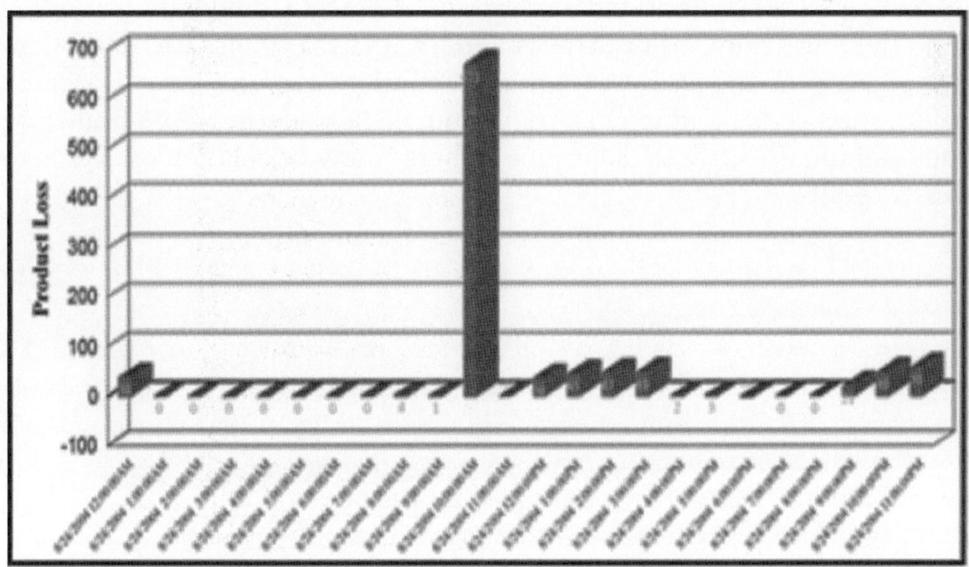

Figure 10.5. Hourly processing variance.

Since all manufacturing control system data are collected in real time and in an open transactional database format, automated collection of real production results by the business systems is possible.

Low-risk Integration with Existing Control Systems

Since this technology does not interfere with the existing control algorithms of control systems, there is minimal risk in implementation. This means that the time it takes to validate the process once the system is in place is minimized as well. Minimizing the amount of integration with the control system also reduces the amount of time required to implement the system as a whole.

Conclusion

The technology described has been successfully implemented in seven manufacturing facilities. These systems have been collecting correlated operational phase data and measured process data continuously, starting immediately upon installation. Some have been collecting data for over 3 years.

The data collected is currently used to generate real-time operational reports that include information based on the control system actions, the actual process measurements, and any manually collected process data. Product loss calculations are currently utilized to generate reports that calculate and display product losses across single process units and logical groups of process units.

The data are being utilized by standard analysis tools, such as Microsoft Excel, to determine the actual energy consumption (e.g., power, gas, steam) per unit of finished goods and to perform various data analyses across multiple process units simultaneously to determine if there is any correlation of events to the measured results.

CHAPTER 11

Closing the Gap: B2MML Driving Successful Integration

Presented at the WBF European Conference, November 13–15, 2006, by

Elspeth O'Brien
Senior Systems Architect
elspeth.obrien@citect.com
Citect
West Midland House, Temple Way
Coleshill, Birmingham, B46 1HH, UK

Abstract

Integration between Enterprise Resource Planning (ERP) and Manufacturing Execution Systems (MES) has been successfully accomplished in the past with custom-made solutions, but this is not a viable long-term solution. Increasingly, ERP and MES vendors are supporting the Business To Manufacturing Markup Language (B2MML) standard. This provides exciting opportunities for companies, as the time invested in designing, implementing, and maintaining these systems can be reduced.

B2MML has provided a vehicle to enable traditionally disparate IT and engineering departments and personnel to communicate. This integration is vital to the success of solutions that reach from the plant floor to the boardroom.

This chapter describes several important lessons learned while implementing these integrated solutions utilizing the B2MML standard—from ensuring B2MML file encoding compatibility, to handling run-time errors in a production system.

Perhaps the most critical lesson learned was the importance of communication—in particular, the need to communicate the interpretation of the standard between project parties. This chapter will illustrate these issues in the context of recent manufacturing projects.

Introduction

Today's manufacturers are working in a global economy with unprecedented global competition. Manufacturers need to produce more for less and must be operationally effective to remain viable. In order to achieve this, a successful company needs to have efficient system integration from the plant floor through to the enterprise business systems.

ERP systems typically handle the manufacturing, logistics, distribution, inventory, shipping, invoicing, and accounting for a company. ERP software has seen great success, and most manufacturing companies have adopted an ERP system. MES have evolved significantly over the last decades, and when coupled effectively with an ERP system, they provide tight integration between the plant floor and business systems. Figure 11.1 depicts the history and future of successful interoperability.

An MES is an operations information system that is responsible for monitoring the execution of plant operations and maintaining data flow to the other systems—primarily the higher-level systems that manage the business. Traditionally there was little connection between the low-level plant floor control system and the high-level company-wide business systems. Today's manufacturers need to be connected to allow them to make decisions based on real-time information. Custom-made integrations are too costly and impractical for the future. B2MML is the most promising emerging solution to realize standards-based integration.

B2MML, a set of Extensible Markup Language (XML) schemas, has been developed by the WBF as a practical implementation of the objects and attributes described in ISA-95 parts 1 and 2. B2MML, now in its third revision, supports companies interested in following ISA-95 for integrating enterprise and business systems, such as ERP, with manufacturing systems, such as control systems and MES.

Using the B2MML standard to integrate MES and ERP systems has numerous benefits. It provides the project with scope, structure, and the ability to succeed with abbreviated project lengths. This chapter describes the various forms the interface can take between MES and ERP systems, identifies project stakeholders, reinforces the importance of communicating the interpretation of the standard, and finally, shares technical integration lessons.

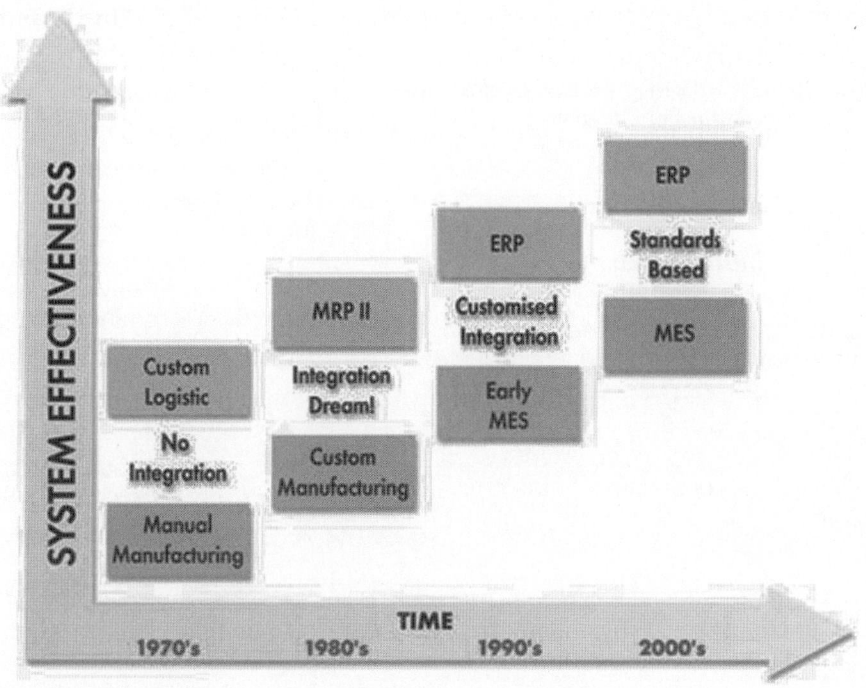

Figure 11.1. Evolution of production to business interoperability.

What Is the Interface?

The B2MML standard applies to the interface between Level 3 and Level 4 of the hierarchical model defined in part 1 of ISA-95. This interface is depicted in Figure 11.2. Level 4 covers business planning and logistics and is typically the domain of ERP and supply chain management systems. MES systems focus on Level 3, which entails the execution of the production process.

Part 3 of the ISA-95 standard describes a large collection of activities related to Manufacturing Operations Management (MOM). Figure 16.3 in Chapter 16 depicts a high-level view of the MOM activities and functions and a possible integration scenario. The activities above the dotted line that starts with an X are the responsibility of Level 4 (i.e., ERP) system, and those below the line are the responsibility of the Level 3 (i.e., MES) system. The figure highlights the large range of integration scenarios that are possible. Determining where to draw the line will depend on a number of factors, which include the following:

- *Legacy systems.* Companies will often have existing systems that will not be decommissioned for the project.

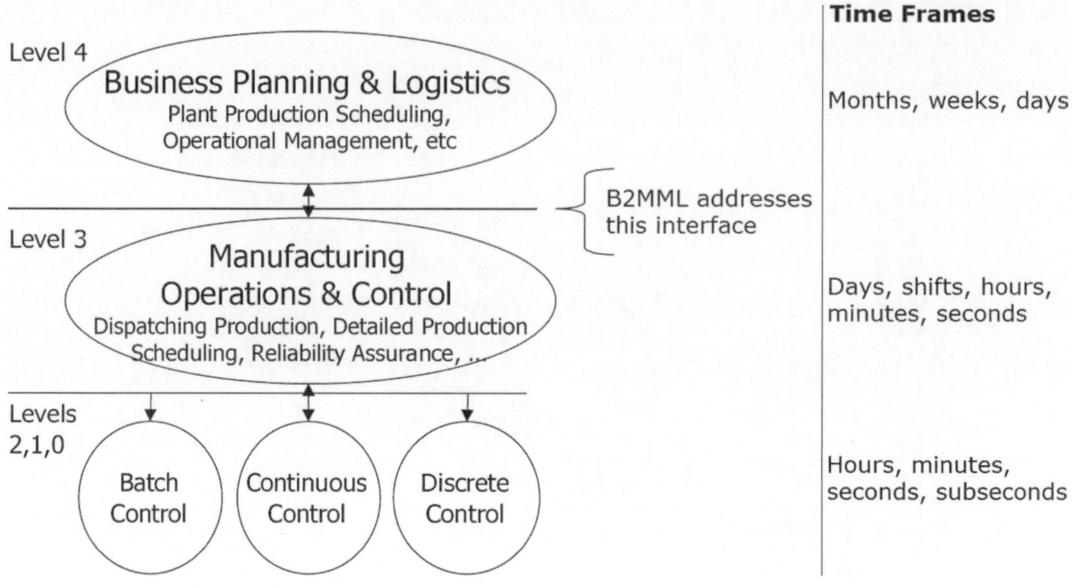

Figure 11.2. Functional hierarchy of activities with characteristic time frames.
Source: Adapted from ISA-95.00.01-2000, figure 3, page 19.

- *Timeliness of data*. Figure 11.2 gives an indication of the typical response times for the system at each level. Consultation with stakeholders will help reveal how often activities need to occur, and therefore, which system will be responsible for supporting that process.

- *Cost and availability of new systems*. The solution will be dependent on the availability and affordability of suitable software in the market.

The ISA-95 standard has provided a large amount of flexibility of responsibilities between Level 3 MES and Level 4 ERP systems. It is therefore an inherent requirement that B2MML be flexible. The B2MML standard is flexible on a number of levels, from which schemas to implement, to which parts of the schema to utilize, to the addition of custom fields, and so on. It is important that the interpretation and implementation of the standard be communicated to the relevant stakeholders.

Identifying the Stakeholders and Project Team

Embarking on an integration project can be daunting and expensive, especially if the integration is part of a much bigger project involving a new ERP or MES. A project team will need to consist of members from a number of different departments within an organization—each with different skill sets and responsibilities. The project team will bring together personnel that don't interact regularly or at all. As stated in part 3 of ISA-95, the arrangement of roles and assignment of roles to personnel or systems will obviously vary between different organizations.[1] It is therefore important that the project team be selected wisely and that all stakeholders are identified.

Identifying all the stakeholders is critical to the completion and ongoing success of the project. Stakeholders should ideally be identified prior to the start of the project. The list of stakeholders will potentially be quite large and will typically include the business project sponsor, IT, human resources, finance, users of the future system, and the development team. Once all the stakeholders have been identified, a communications plan should be formulated.

Communications planning is determining how to communicate important messages to stakeholders in the most effective way possible. The result of this planning is called a communications plan. The communications plan will identify all the stakeholders, the information they require, how frequently they require it, and in what format (e.g., e-mail, presentation, meeting). It is important that the communications plan is used and updated as needed during the project. For example, the project may require a new company or person to be contracted, and they will need to be added to the plan.

The communications plan and the identification of stakeholders have an important role for several aspects of the project. Importantly, it aims to identify all the future users of the system, thus allowing the company to construct a change management plan, promote the project, and identify training needs. Once the interface between the MES and ERP has been defined and the stakeholders have been identified, the resulting communications plan should be utilized to communicate the interpretation of the standard to all relevant stakeholders.

Interpreting and Communicating the Standard

B2MML consists of nine schemas and a common schema. Version 3 of the B2MML standard also includes an extensions schema. The nine main schemas map to the nine object models defined in ISA-95 parts 1 and 2. Prior to selecting which schemas to use and specifying the B2MML message details, a company will need to determine the interface between their Level 3 MES and Level 4 ERP. Once this interface is known, the schemas that will be utilized can be chosen. The requirements of the business will be core to the design of the interface.

The business requirements, the current business position, and the company culture will help to dictate the scope of the interface. The schemas provide the ability for customers to have a partial, phased, or full implementation (i.e., the ability to utilize one or more B2MML schemas). This flexibility is beneficial for customers for a number of reasons, including the following:

- The lessons and confidence gained from implementation of a small number of schemas can potentially shorten future projects. The partial implementation serves as a proof of concept and can shorten the initial project length, allowing prioritized business requirements to be realized earlier.

- It may be too costly to implement the complete interface up front. Each schema needs to be fully scoped and specified, tested and commissioned. The corresponding training load and change management may lead a company to adopt a partial or phased implementation.

- Implementing the full interface may introduce too much up-front complexity. A larger number of stakeholders will need to be engaged to implement all the schemas (e.g., Maintenance personnel will need to be involved in the Maintenance schema but perhaps not the Production Performance schema).

Once the schemas have been identified, the details of how each schema is going to be utilized need to be determined. Each schema allows for substantial customization, and therefore it is essential that each instance of the message be identified. Additionally, the timing and frequency of each message instance should be documented. For example, a company has identified that it wants to forward finished goods information from the MES to the ERP each time a pallet is received in the warehouse. Figure 11.3 highlights a section of the Production Performance schema that could be used for this purpose. This is an example of a Production Performance message instance.

The documentation for each message instance should describe each field in detail. The description of a field should include, at minimum, a good description, data type, maximum length, and an example value. Custom fields can be especially useful, and it is important that they are documented well. Custom fields, also known as user element extensibility, allow customer-specific information to be communicated between Level 3 and Level 4 systems. Custom fields are particularly useful for companies that have assigned particular responsibilities to one system; however the other system may also need the output of that responsibility. Instead of reproducing the business logic or master data, custom fields allow the required information to be passed between Level 3 and Level 4 systems.

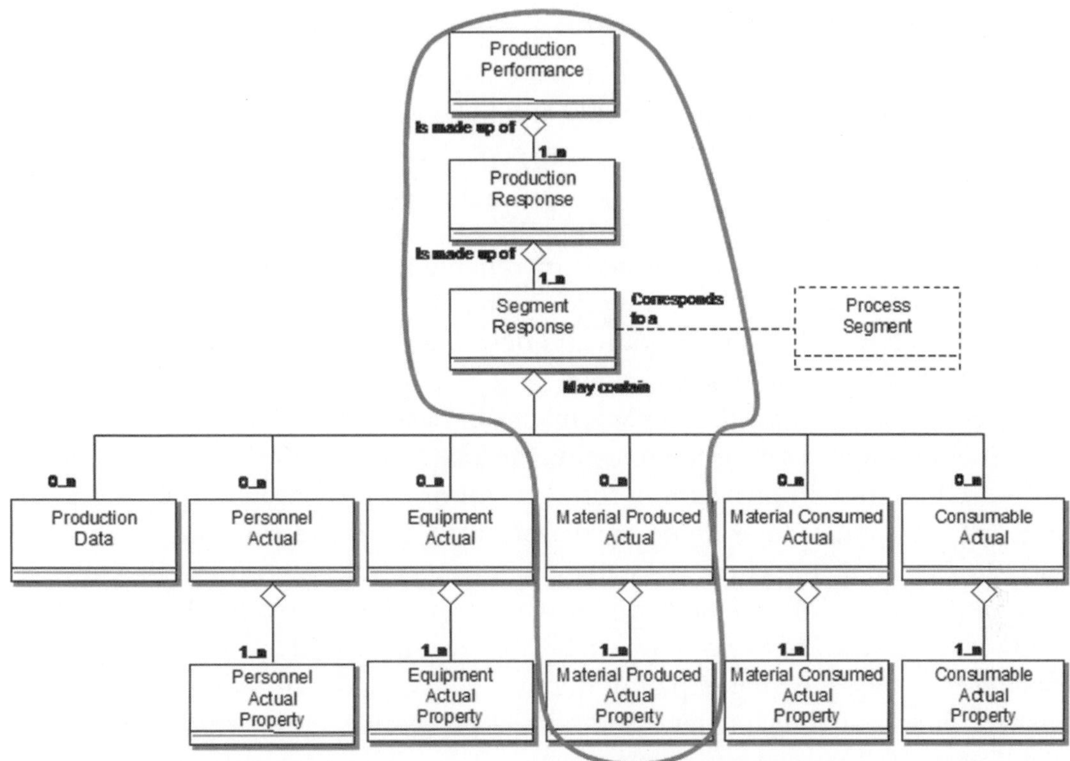

Figure 11.3. Model of exchanged production capability information.
Source: Production performance schema documentation, version 2, September 23, 2003.

The importance of documenting and communicating the use of the standard is critical, especially when engaging contractors for parts of the project. Misunderstandings can lead to broken interfaces, erroneous data, frustration, and wasted time. Perhaps one of the most critical parts of the project is establishing all relevant stakeholders. Often, the appropriate documentation has been created but has, unfortunately, not been distributed to all the required stakeholders.

Technical Integration Lessons

During the course of implementing projects utilizing the B2MML standard, we have learned some interesting lessons. Even with the best laid plans and good project management, testing and commissioning a project can uncover some interesting challenges. The impact of small issues can be large if the right processes are not in place to allow for a quick resolution. This section discusses a few of the lessons learned while commissioning integration projects.

Runtime—Keeping the System Alive

The real test of an integration project is the day that it "goes live." The MES and ERP systems will need to work reliably and in coordination with each other. Even the best-tested systems will reveal unexpected and undesirable situations that were not identified earlier. It is important that management allows the plant to run at a reduced production rate to smooth out any teething issues before ramping up to full production. This time is invaluable to allow staff to adopt the new working practices and to prevent work-arounds from becoming permanent. Here, we present a typical implementation and describe areas that will require attention during the "go live" phase and beyond.

The journey a B2MML message makes between the MES and ERP systems often involves a number of hops. Unlike a closed-loop control system, messages are typically sent without waiting for a receipt or confirmation of delivery to the MES or ERP. In order to examine the reasons for this architecture and to plan support for such an architecture, we will look at an example. Figure 11.4 displays the journey that a B2MML message might make from the MES to the ERP. The MES produces a Production Performance B2MML message based on an event and publishes it to a message queue that guarantees delivery. The message is placed in a First In, First Out (FIFO) queue and awaits processing by the middleware. The middleware will process messages from the message queue and update the ERP accordingly. Typically, the middleware will create transactions for the ERP to batch process periodically. It is important that messages be processed in strict order to prevent situations, for example, where a production cancellation gets processed before the corresponding production declaration.

The architecture shown in Figure 11.4 is common for a number of reasons. First, one of the main reasons is that the manufacturer's current ERP does not support B2MML. The middleware provides the required translations between the MES and ERP. Secondly, plant and corporate networks have historically been disconnected. The message queue software or middleware can provide the necessary secure bridge across the networks. Thirdly, the MES is operating in real time and needs to be available at all times. It is not uncommon for an ERP system to be down for several hours each month for maintenance; therefore it is not feasible

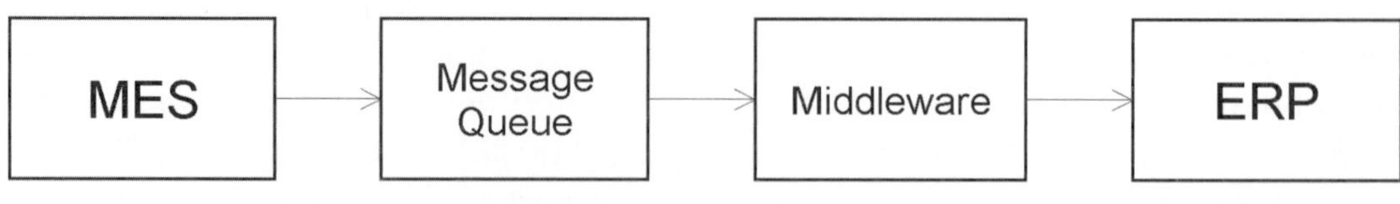

Figure 11.4. Example system architecture.

for the MES to wait for responses to messages if the plant needs to operate during maintenance periods. The message queue software or middleware can provide the appropriate buffers for messages during this time.

Given the architecture in Figure 11.4, it is important that the systems run on robust and redundant hardware. Similarly, fault tolerance needs to be built into the software systems. If we focus on the reliability of the integration, then there need to be plans to handle failure at each system that the B2MML message encounters. A good design will have answers to questions such as the following:

- What will the MES do if the message queue is unavailable when trying to export a B2MML message?

- How will the middleware handle a corrupt B2MML message? (A corrupt message could be an empty file, be incorrectly coded, have an invalid XML field, and so on.)

- How long can the ERP be down for? In other words, how large are the message buffers and how long before data are lost?

- How does the ERP respond to and report issues while processing a B2MML message?

Answers to these questions may be resolved by implementing extra logic or workflows within the appropriate system. However, when an error occurs it is often because of an unanticipated scenario. In this situation, it is unlikely that the system will be able to resolve the issue itself. It is therefore important that when a failure occurs, the right personnel are notified and the systems continue to function. The right personnel need to have the capability or contacts within the organization to resolve the issue in a timely fashion. The following paragraph describes a scenario where the B2MML message does not reach the ERP.

A forklift driver picks up finished goods from the end of the production line and scans the label on the pallet. The forklift driver is expecting the ERP to provide details of where to place the pallet. The ERP does not recognize the pallet, and the forklift driver has to put the pallet aside or stop moving further pallets until the issue is resolved. This situation can arise when a Production Performance message declares that the finished product has not been received or processed by the ERP. The forklift driver notifies his supervisor, who contacts a member of the project team during commissioning or perhaps the help desk. It is very quick to see that without the right processes in place, valuable production time can be lost trying to pinpoint the issue. The problem can be exacerbated if it occurs outside regular work hours and IT personnel need to be contacted.

The aforementioned scenario is one way of confirming that the B2MML messages are reaching the ERP (i.e., the ERP has knowledge of the finished product and therefore the ERP must have received the appropriate message). Another method is to reconcile information between the MES and ERP. For example, the total production for a shop order or batch in the MES should match the total receipted goods in the ERP. It is important to note that there are many different ways to locate an issue. The personnel on the plant floor will quite often not have access to the middleware and IT systems and therefore will diagnose issues using methods such as reconciliation before escalating the issue.

A holistic approach needs to be taken when commissioning and supporting the integration between MES and ERP. Each system needs to be reviewed in isolation and in conjunction with the whole system. Personnel throughout the organization need to be equipped with the right information and knowledge of how to escalate issues before they impact on production and ultimately profit. In the next two sections, we look at two issues that prevented B2MML messages from reaching the ERP.

B2MML File Encoding

The time from project start to project completion can often span years for a large organization. This lengthy time frame can have implications for the chosen technologies and standards. In one such project, the B2MML standard was in its infancy at project conception, and the chosen ERP did not directly support the B2MML standard. In order to utilize the B2MML standard, middleware was used to translate the B2MML files to and from the ERP. The architecture was similar to that shown in Figure 11.4. This integration presented an interesting issue with encoding.

The encoding for each B2MML message is declared on the first line, as shown in Figure 11.5. The encoding declaration is used by XML processors to successfully decode the file. The Unicode Standard assigns a code point (i.e., a number) to each character in every supported script. UTF-8 is an 8-bit lossless encoding that represents each code point as a sequence of one to four bytes. The default encoding for B2MML is UTF-8.

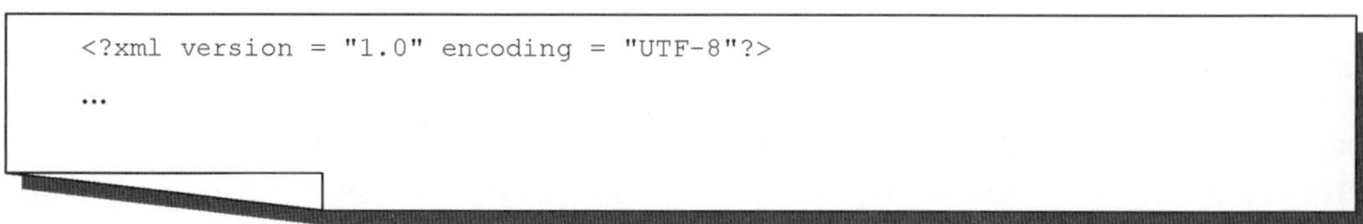

```
<?xml version = "1.0" encoding = "UTF-8"?>

...
```

Figure 11.5. First line of all B2MML schemas.

In addition to the encoding declaration on the first line, some applications insert a particular combination of bytes at the beginning of a file to indicate that the text contained in the file is Unicode. This combination of bytes is known as a signature or Byte Order Mark.[2] The UTF-8 signature in hexadecimal is "EF BB BF."

Support for the UTF-8 signature is not universal. In this project, the middleware did not support the signature, and the file was unable to be decoded correctly. This initially resulted in the entire message queue being blocked during testing. The issue was resolved by configuring the MES to not prefix a signature to the B2MML file and by updating the middleware so that corrupt messages did not block the queue.

Date and Time Fields

It is not uncommon for ERP and MES servers to be in disparate geographical locations. This introduces the interesting complication of dealing with different time zones. This is a good example of where the interpretation of the standard needs to be communicated. In one project, the Level 3 system formatted the start-time and end-time fields according to the documented required format. It transpired that the Level 3 system was using Coordinated Universal Time (UTC) time and the Level 4 system was expecting local time. The situation was easily remedied by updating the Extensible Stylesheet Language Transformations (XSLT)[3] in the MES to use local time.

Conclusion

For manufacturers to remain viable in an increasingly competitive environment, they need visibility of their operations in real time. In order to realize these goals, companies need to have well-connected systems from the plant floor to the enterprise. The ISA-95 and B2MML standards are emerging as vital tools for manufacturers, allowing MES and ERP systems to be integrated successfully.

Adopting the B2MML standard for an integration project has enormous benefits for the initial project and beyond. Manufacturers can approach vendors with clear and specific requirements using a common language: ISA-95. Manufacturers can learn and benefit from the experiences of a wide community that designs and uses the standards. Adopting a standards-based approach to integration also provides the ability to upgrade systems in the future without the headache of custom interfaces.

Major ERP and MES vendors now support the B2MML standard. This provides exciting new options for manufacturers looking to upgrade or integrate their

MES and ERP systems. The best results for integration can be achieved when the project is well managed. A successful project requires manufacturers to identify all stakeholders, create a living communications plan, and use the plan to communicate the interpretation of the standard.

Notes

1. See ANSI/ISA-95.00.03-2005, page 30, section 6.2.
2. For a discussion of display problems caused by the UTF-8 Byte Order Mark, see the W3C Architecture Web site: http://www.w3.org/International/questions/qa-utf8-bom.
3. XSLT is a language for transforming XML documents into other XML documents.

ISA-95 Integration between SAP R/3 and Batch in Pharmaceutical Applications

Presented at the WBF European Conference, October 11–13, 2004, by

Stephan Van Dijck
Consultant Supply Chain Integration
svandijck@alltel.net
ABB
31 Stratford Road
Hudson, OH 44236, USA

Abstract

In this chapter we will share past and present experiences using the ISA-95 integration model for business process transactions and data integration for supply chains. In particular, we will explore an Enterprise Resource Planning (ERP) and batch integration example from a functional, technical, and evolutionary perspective, focusing on regulated industries.

The ISA-95 integration model is used to structure the integration processes between business systems and the plant floor. Through the use of the ISA-95 structures, a "common denominator" business model was established for integrating SAP's R/3 PP-PI transactions and data structures with an ISA-88 batch automation data structure. The integration is transactional, bidirectional, and addresses fault tolerance for long-running, asynchronous transactions between SAP and the plant floor. In particular, we will discuss the challenges of integrating SAP PP-PI and a batch solution and how these challenges are addressed. We will share and compare our experiences in working with Microsoft .NET and IBM WebSphere products with these types of integration projects.

The setting for the applications presented will be the pharmaceutical and food and beverage industry—specifically, our experiences in deploying a consistent security model, business processing, data transformation, master data, and abnormal situation handling will be discussed.

Introduction: Supply Chain Perspective

In regulated industries such as pharmaceuticals or food and beverage, the role of the supply chain is a key contributor to the bottom line profitability of an organization. Making sure the value chain is well integrated remains a prime focus for investment.

The following points cover five areas of manufacturing and distribution in which companies will have to concentrate their efforts if they are to create a supply chain that fulfills their future requirements:

- Supply chain or demand synchronization and strategic sourcing
- Scientific manufacturing
- New product and process development
- Restructuring and asset rationalization
- Techniques for extending reach to the customer

The first objective is to increase productivity and maximize the efficiency of a company's existing manufacturing facilities. The second is to ensure that its supply chain conforms to the European Union's (EU) and U.S. Food and Drug Administration's (FDA) new compliance agenda. The third objective is to prepare and reorganize its plant(s), processes, and people for the future (Fig. 12.1).

Figure 12.1 describes the market pressures companies are faced with: market changes, regulatory pressure, and fast product evolution. Factors that are available for companies to handle these pressures include the following:

- *Portfolio.* All companies have limited resources, so it is imperative to direct those resources—be they capital, plant, skills, or management time—to the areas of the business that create most value. Many pharmaceutical firms tend to have limited portfolios because they look backward, not forward. They have supply chains that are engineered to manufacture traditional and well-established products, rather than a much wider range of products, many of them biological rather

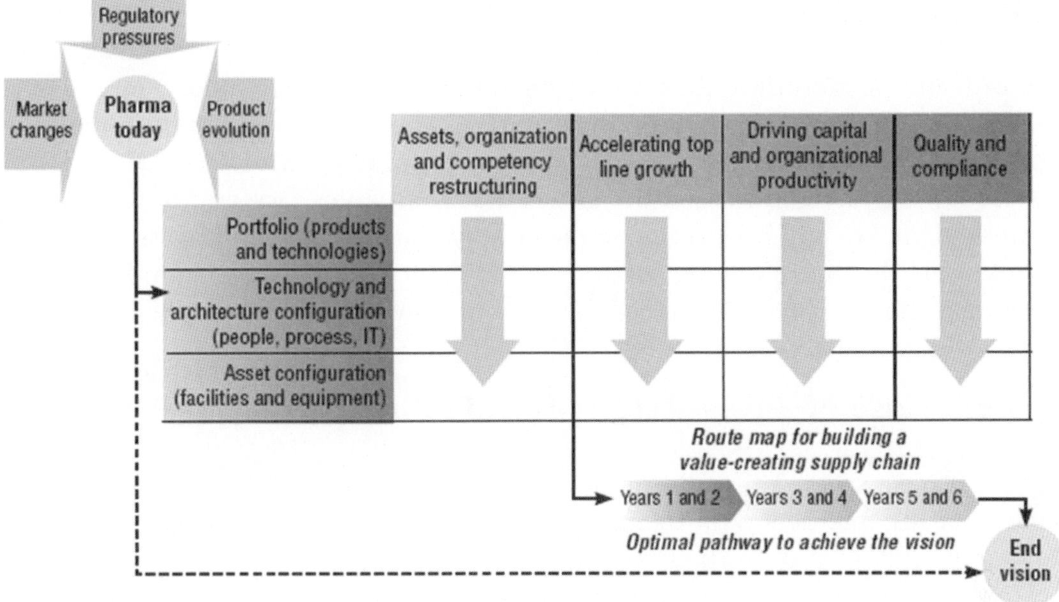

Figure 12.1. Effects of market pressures.
Source: IBM Business Consulting Services

than chemical in nature. They will ultimately have to reorganize their manufacturing assets and, in doing so, take a zero-based view of the business, since a piecemeal or step-by-step approach to the restructuring of an organization and its assets is not effective.

- *Evolving technology.* Companies should seek to evolve the entire spectrum of their automation technology, from control systems to ERP systems, including the middle Collaborative Production Management (CPM) tier.

- *Asset configuration.* Companies must learn how to make better use of their plants and people and reduce their fixed costs. People, equipment, experience, and intellectual property must all be deployed in a more adaptive and flexible way. Three elements are essential here: a Lean manufacturing culture, long-range modeling, and an integrated supply network. Some of these elements are hard to implement. For example, Lean manufacturing may be constrained by regulatory boundaries, but it is worth exploring to improve flexibility.

These factors enable a company to restructure its assets, accelerate growth, and drive productivity and quality to a competitive bottom line.

Role of ISA-95

The previous section describes some general drivers for regulated companies to improve performance in a competitive market. These drivers have been mapped against the ISA-95 activity models, as illustrated in Table 12.1.

So far, we have addressed the issues that companies face and how they can address them. We will now focus on the supply chain as a major means of achieving the goal. In the following section, we will provide specific examples of the functional challenges that have been faced and how they were solved.

Typical Case of a Regulated Manufacturer

Figure 12.2 is a schematic representation of a typical production process. The ERP level typically interacts with the plant in three areas: material management, production planning, and production data.

Material management makes sure that the right materials come to the floor at the right time, in the right quantity. It is here that Lean manufacturing could be explored. Discrete manufactures have been using the principles of Lean manufacturing for a long time, yet in regulated businesses, these principles have seldom been implemented.

Production planning, which drives shorter time-horizon scheduling, is often well established. Typically, the potential benefit of integration toward the operational side is underexploited.

Table 12.1. Performance improvement drivers and ISA-95	
Area of focus	*Corresponding ISA-95 activities*
Demand synchronization and strategic sourcing	Order processing, scheduling, and procurement
Restructuring and asset rationalization	Material and energy control, procurement, production control, maintenance management, product inventory control
Scientific manufacturing	Production control
New product and process development	Process development relies in part on feedback from manufacturing, in an attempt to improve and adapt to change. Research and Development (R&D) for new products is not directly part of ISA-95, but the interaction between R&D and manufacturing is recognized and included.
Techniques for extending the reach to the customer	Marketing and sales, dynamic scheduling

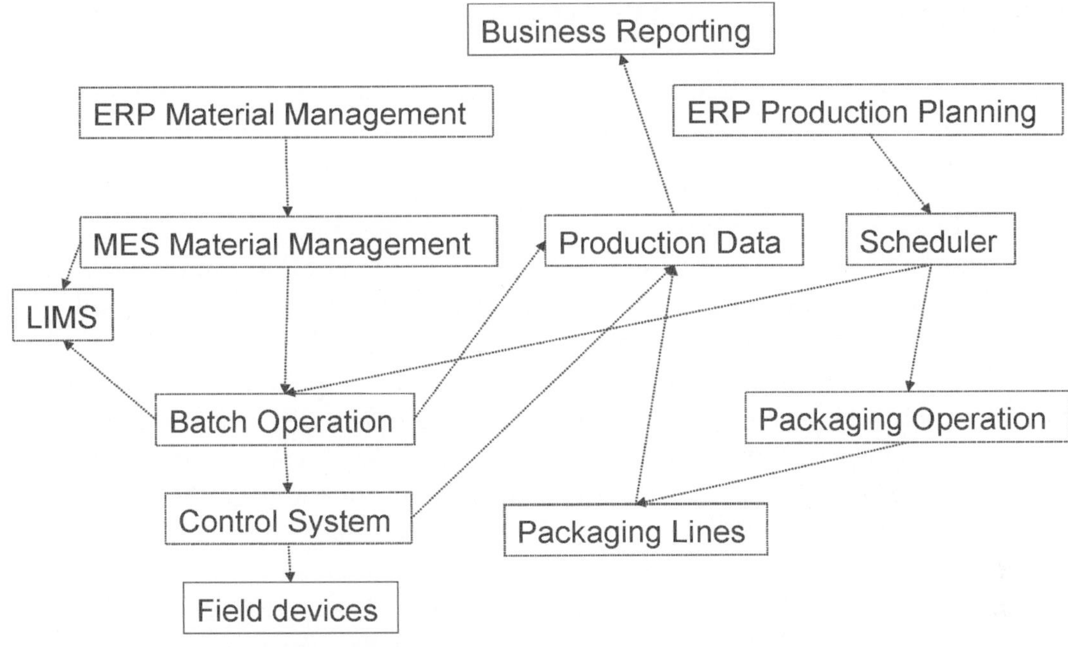

Figure 12.2. Typical production process.

Production Response ensures that information on production runs is collected from real-time operations and that values are transferred up to the business system. Again, the level of integration is inadequate. A tighter integration would enable business systems to react to disturbances more quickly. An analogy would be the type of closed-loop control that process control has evolved. Given these requirements, there are a number of functional challenges in implementing a process flow.

Functional Challenges

Security

Any integrated system consists of multiple software packages. Typically, these have their own security model, particularly when they are hosted on multiple platforms. In many present-day scenarios, an operator may even have two screens, one for ERP information and the other for process or control information.

When programs call or pass information to other programs, the same security concerns apply. WebServices, the technique all major vendors are promoting, is an excellent example of a technology that requires the careful consideration of security issues by both the service provider and service consumers. Irrespective of the approach, a synchronized mechanism for managing interapplication security must exist.

Business Process Synchronization

Business processes in most companies change constantly. Mergers and acquisitions, new products, new raw materials, new technologies, new regulations—these factors all contribute to changes in business processes. Companies that can quickly adapt their business processes will be able to turn change into a competitive advantage.

Through intracompany optimization, companies have engineered their business practices to enhance overall performance. By implementing internal system solutions such as ERPs and supply chain planning and execution systems, company management can make informed business decisions. Intracompany optimization extends real-time information throughout the organization, ensuring synergy among operations, finance, sales, purchasing, and customer service.

Business process synchronization takes supply chain integration to the next level of efficiency, utilizing standardized information formats and communication points between organizations. Business process synchronization eliminates costs associated with inefficient movement of goods, redundant processes, and excess inventory. It also promotes the collaboration of all supply chain partners—suppliers, manufacturers, distributors, wholesalers, third-party providers, transportation companies, and retailers.

Data Transformation

Today, most supply chain partners do not share business practice knowledge or content. Not only is the lack of standardized technology an impediment, the information itself can also be an obstacle. When information is confidential, transmission requires a very secure data communications connection. If the data are volatile or a database is too large to transmit, then continual, real-time access may not be feasible. Data mapping between systems must also be addressed. An excellent example is product codes, which are often different between ERP and manufacturing systems. Last but not least, data formats differ from system to system. A classic example is SAP's date format, which is different from what most other programs use.

Master Data

All systems have their own data models and structures. The set of data that defines the master data is likely to be needed in multiple systems. Each system has slightly different content or usage, entitling it to be the "master" of the data structure. The problem becomes more complex when changes are necessary to the master data. The change must be made in each system and must be synchronized with potential local changes. This opens up a potential for data inconsistency, errors, and bad quality.

Abnormal Situation Handling

Occasionally, things do not go according to plan. The way these abnormal situations are handled is critical to developing trust in an integrated system. If, for example, a fermentation vessel fails and goes out of production, then the production orders currently scheduled for the vessel need to be rescheduled. Business rules specify which customers may have to be notified if the delivery dates change. It is our experience that the careful handling of "what ifs" can be as complex as an entire "normal" situation project.

Solution

There is no magic bullet for implementing an integrated supply chain solution. Luckily, the issue is well recognized, and technology is quickly moving to help mitigate some of the issues. Later on in this chapter, we will discuss two platforms that help address this: .NET from Microsoft and WebSphere from IBM.

In general, we have used off-the-shelf products and avoided custom programming. Support, upgrading, and trained resources are the most obvious reasons for this. The decisions have been driven by a few key requirements:

- *Transactional integrity.* A transaction is a coordinated series of modifications of data (e.g., data stored in a database or a file system), guaranteed either to be successfully executed in its entirety or not to be executed at all. To implement a transaction, a record is kept of the state of the data store before the transaction begins, and if one of the modifications fails, then the transaction returns failure and the initial state is restored (or "rolled back"). Transactions are used to maintain data integrity, and consequently, play an important role in business software programming.

- *Persistence.* Writing data to a database in the context of the transaction is a key requirement.

- *Business process enabled.* Middleware has evolved from data handling to business process handling. The market is moving rapidly to provide solutions that allow more flexible business processing.

Extending the use of the data-rich middleware solutions toward flexible decision making is becoming a strong requirement.

- *Business intelligence capability.* The best business intelligence and business analysis applications should be provided.

- *Visualization.* Personalized thin clients in a distributed environment that includes PCs, handhelds, and mobile phones should be used.

- *Engineering tools.* Each package involved may have its own engineering tools. The less tools, or the better they are integrated, the more effective the implementation and subsequent maintenance will be.

ISA-95 will play an increasingly important role in each solution. It is the "common denominator" data structure that the industry agrees upon that, hopefully, in the future more people will adhere to. Just as ISA-88 has done, ISA-95 makes the integration easier to implement by providing a common vocabulary and a set of models.

Solution Technologies

In our implementation work, we have focused on two main technology platforms: .NET and Java. Each has its own specifics, and we will try to highlight a few that we feel are important. The Microsoft .NET solution stack has been evolving over the past few years. Microsoft has gained a lot of momentum, and BizTalk Server 2004 is definitely aiming at the integration space in industrial settings. We have focused on three areas: integration, analytics, and collaboration. The Microsoft technology stack deployed is described in Figure 12.3.

Earlier in the chapter, we listed functional requirements. Table 12.2 maps the features of the .NET platform to it. A similar technology stack (Fig. 12.4) and table (Table 12.3) is shown for IBM's WebSphere solution. The IBM technology stack is described in Figure 12.4. Table 12.3 maps the features of the IBM WebSphere platform.

Resulting Solution

Both .NET and Java enable a solution for integrating the supply chain. Some of the relevant results are as follows, focusing on work done linking SAP R/3 PP-PI to batch control:

- *Security.* It is possible to implement a unified solution with each platform. Specifically, Microsoft's Single-Sign-On utility provides an excellent environment, where one security model can be used with

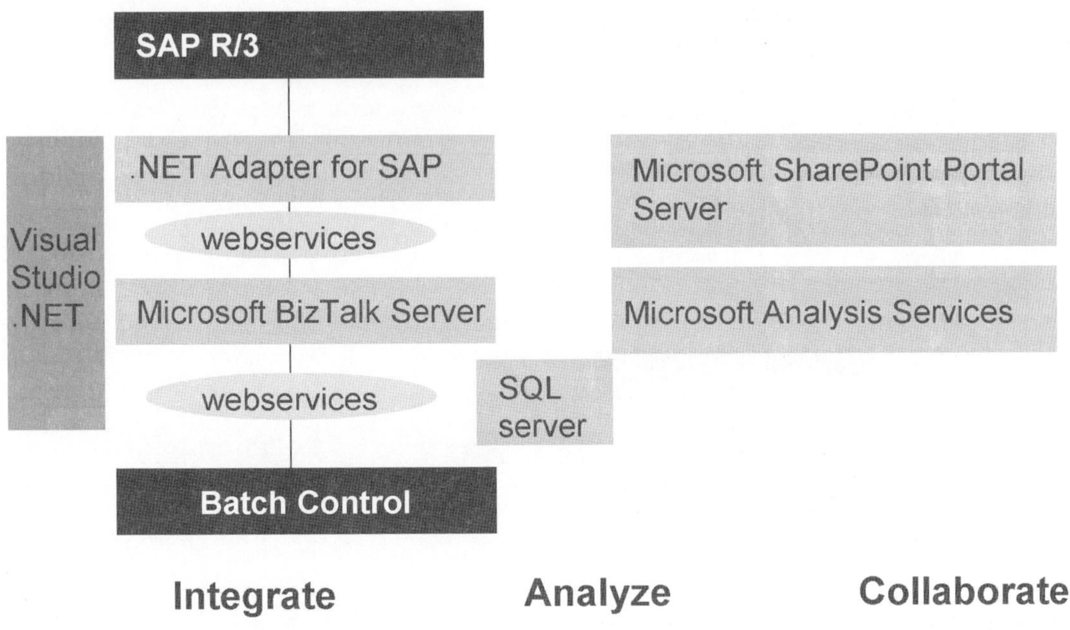

Figure 12.3. Microsoft technology stack.

Table 12.2. Functional requirements using the .NET platform	
Functional requirement	*.NET feature*
Transactional integrity	BizTalk Server 2004 has an extensive model for state handling of transactions and handles roll-back situations.
Persistence	SQL Server is used to write data while the combination of SQL Server and the tight integration with BizTalk makes sure that the context of data is stored correctly.
Business process enabled	Orchestrations are the concept used by BizTalk to implement the business logic in a graphical format. In addition, business rules can be used to make the implementation of decision logic more accessible to a business user.
Business intelligence capability	SQL Server comes with an Analysis Server, used for business analytics.
Visualization—Web enabled	Share Point Portal server is integrated well with BizTalk Server and the office suite products, making it the technology of choice for distributed visualization.
Engineering tools—integrated and relatively easy to use	Visual Studio .NET provides one integrated tool for all applications.

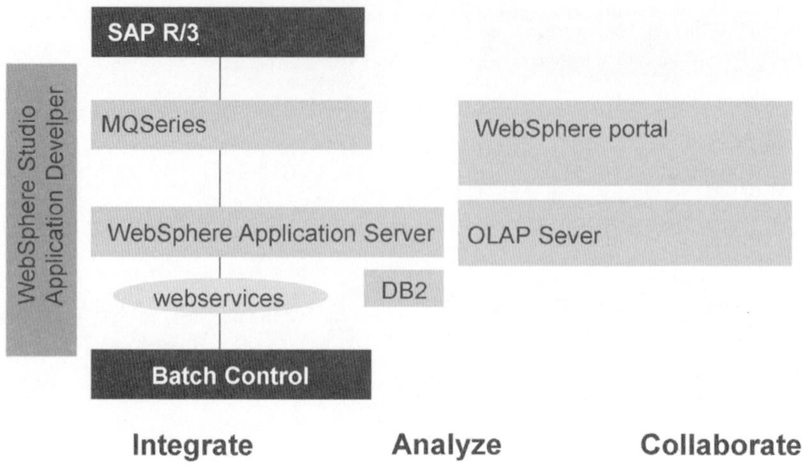

Figure 12.4. IBM WebSphere technology stack.

Table 12.3. Functional requirements using the IBM WebSphere platform	
Functional requirement	*IBM WebSphere feature*
Transactional integrity	WebSphere's J2EE Application Server and its implementation of Java beans has an extensive model for the state handling of transactions. Handling of roll-back is typically left to the implementer.
Persistence	Java beans can write to many databases (e.g., DB2, Oracle, SQL Server). This provides an open environment that is capable of implementing complex structures. The business transaction context relationship is harder to establish.
Business process enabled	WebSphere comes with industrial solutions in a supply chain that can be used as a starting point and altered to match company-specific requirements. Business Process Management (BPM) with IBM WebSphere MQ Workflow is another solution to automate business processes.
Business intelligence capability	DB2 has Online Analytical Processing (OLAP) functionality built in.
Visualization—Web enabled	WebSphere portal delivers a single point of personalized interaction with applications, content, processes, and people for a unified user experience.
Engineering tools—integrated and relatively easy to use	WebSphere Studio Application Developer is a family of integrated development environments for Web, Java, J2EE, Web Services, Extensible Markup Language (XML), and data applications. It is a key element of the IBM Software Development Platform used to automate and integrate the software development process.

multiple integrated applications (providing that similar hardware
and operating systems are used).

- *Transactional integrity.* The PI-PCS interface has a strong transactional
  behavior built in. Downloaded information is only complete when an
  acknowledgement signal is obtained from the receiving system. Solutions such as BizTalk Server and WebSphere permit the processing of
  this transactional state cycle quite adequately.

- *Interoperability.* In the R/3 environment, PP-PI's PI-PCS interface is
  not available as a WebService. By deploying the techniques mentioned, we were able to expose the PI-PCS interface as a WebService.
  BizTalk and WebSphere were able to interact with these WebServices
  successfully.

- *Data Mapping.* The transformation of data structures is a particular
  concern, going from relational tables to hierarchical XML structures
  and back. Also, the data semantics change. The solution obtained
  complied with both requirements.

Interoperability: Comparison of .NET and WebSphere

Earlier in this chapter, both technology stacks were described: .NET and Web-
Sphere. In Table 12.4, we will explore both platforms. A lot has been written about
.NET and Java, and a comparison often leads to very animated discussions. We
have tried to speak from our experience in supply chain integration. Four stars is
the highest rating in Table 12.4.

Conclusion

Focus continues to be put on the supply chains and how they can generate value
for companies. Success in integrating these supply chains within an organization
or across multiple organizations can be an important means of achieving the goal
of growth and success.

ISA-95's integration concept fits nicely with the vision of integrated supply
chains, by addressing the major issue of business to manufacturing information
exchange. A cross-industry adoption of the standard will further accelerate the
success of integration efforts.

Challenges such as security, transactional integrity, and persistence of data
have been addressed in multiple competing platforms—Microsoft's .NET and the

Table 12.4. J2EE versus .NET			
Criteria	*J2EE*	*.NET*	*Comment*
Development environment tools	**	****	Visual Studio .NET is the single tool for the .NET platform. WebSphere Studio covers a lot of ground, but there are still other applications that require their own tool.
Scalability	***	**	Java code executes on mainframe computers. WebSphere Server can easily be distributed. BizTalk Server and SQL Server support scaling too, but IBM delivers successfully to small and large applications, and the evidence in scalability is still there.
Platform dependency	****	*	Java is platform independent. .NET runs on the Microsoft platform only.
Performance	***	***	In our experience, both solutions provide adequate performance.
Speed of development— code generation	***	**	Both environments make use of Wizards to generate code. Wizards are excellent, but often one wants control over their output. Java, more open, is more forgiving and lets the developer modify the generated code.
Open standards	****	*	By definition, Java is an open standard.
One function, one software	*	****	In the IBM portfolio, there are many different, sometimes overlapping solutions to the same problem. Microsoft's solution portfolio is more streamlined. One solution for each challenge makes the choices easier, but also reduces the flexibility gained from a wider choice of solutions for the same problem.

IBM WebSphere product family being the dominant players. Both solutions have advantages and shortcomings, but their interoperability is largely enhanced by the use of WebServices, enabling the best of both worlds for the end user.

Supply chain integration will always be a challenging endeavor. Accepting the challenge is the starting point on the road to success.

CHAPTER 13

Setting the Standard for ERP and MES Integration

Presented at the WBF
North American Conference,
May 15–18, 2005, by

David Cornell
Systems Manager
cornell.dl@pg.com
Procter & Gamble
8256 Union Center Blvd.
West Chester, OH 45069, USA

Abstract

The Procter & Gamble (P&G) Manufacturing Execution System (MES) organization has implemented Business To Manufacturing Markup Language (B2MML) with SAP and several MES and batch control systems. We worked closely with the WBF B2MML working group to coordinate the use of B2MML for common information exchange across many companies. We worked with internal development teams of Enterprise Resource Planning (ERP) and MES vendors to guide their development of B2MML interfaces in their products. Lead site implementations of B2MML are in progress in most P&G business units. Significant value will be delivered by the implementation of a single industry standard MES-ERP interface in our global company with many manufacturing systems.

At P&G, we developed a B2MML interface for SAP that exchanges information on orders, production, consumption, and quality of material produced and consumed. These B2MML messages are used by two MES systems: one internally developed and one from GE Fanuc who worked with P&G, SAP, and the WBF working group to develop their interface. P&G continues to work with SAP on their interface development. We learned how SAP interacted with B2MML so that we could design our own interface for SAP, to be replaced with SAP's when it is delivered.

Introduction

Tightly integrating our ERP and MES systems has become the focus of many projects aimed at creating value at P&G. The focus on external information exchange with customers has been a priority for a long time. Now, to achieve the cost savings and improvement in the supply chain, information must be exchanged accurately and much more quickly among all information systems within our company. This includes the information exchange between the ERP and MES systems, which is the focus of the ISA-95 standard. B2MML, the Extensible Markup Language (XML) implementation of ISA-95 developed by the WBF, is becoming more widely accepted.

Significant progress has been made at P&G with our deployment of the ISA-95 standard via B2MML. Our pilot test of a B2MML Production Schedule download from ERP and Production Performance upload to ERP, with our Fabric and Home Care Business's internally developed MES system, succeeded. Further deployment has begun. A prototype in our Family Care business of the same Production Schedule download into our key MES application also succeeded. It, too, is expanding in that business.

More importantly, our key ERP vendor has embraced the ISA-95 standard and B2MML. They initiated a series of working group meetings to solicit customer and MES vendor input to their ISA-95 development. P&G has participated and looks forward to the results. Our key MES vendor has begun development of an XML ISA-95 interface to its MES application. P&G will be prototyping this interface.

P&G is now expanding the scope of our ISA-95 implementation by adding new messages and including other business units. In 2005, two prototypes in our Health Care and Beauty Care businesses will test this expanded ISA-95 scope, which now includes Quality Test Results related to Material Production.

These two prototypes will also test a newly developed ISA-88 structure in our key MES system, with its batch history archive modules.

Vendors Develop Standards-based Interfaces

ISA-95 in ERP

Responding to requests from many of their users, our ERP vendor has embraced the ISA-95 standard. They solicited input on this development from customers and vendors of information systems that exchange information with them. P&G is a participating member of these workshops.

Our ERP vendor is currently developing a B2MML interface that will become part of their ERP product. They will use their middleware application to map to B2MML. A first version of some of their B2MML has been released and is being reviewed at P&G.

ISA-95 in MES

Our key MES system vendor will be adding ISA-95 integration to their MES product by adding a middleware application packaged with the product. During the prototyping phase, P&G will work with them on the design and development of the mapping between their data structures and our ISA-95 XML structure. As the B2MML structure for some parts of the data P&G intends to exchange has not been firmly established, the XML mappings will be P&G specific. However, the foundation will be laid for the development of standard B2MML mappings that all customers can use.

The scope of the prototype implementation includes processing inbound Production Schedule B2MML into the MES Production Plan and new Bill of Materials structures. The scope also includes generating outbound Production Performance B2MML that maps Production Event Status to Order Confirmation, Material Lot Production Events to Material Production, and Material Lot Genealogy to Material Consumption. Plus, a new Test Performance XML, created by P&G as a first draft of a new B2MML schema, will be generated containing Quality data.

ISA-88 in MES

Our key MES vendor is also utilizing ISA-88 to store batch history information. These new structures will hold the Batch, Unit Procedure, Operation, and Phase events, plus their parameters. This will improve an earlier version of batch history to now allow storing nearly all types of batch data in a manner that is consistent with ISA-88. Many of the manufacturing operations at P&G that require close integration with ERP are batch processes.

Application of Integration Standards at P&G

Value of Standard Integration

The inherent value of a standard-based integration mechanism seems self-evident yet bears quantifying. The lower cost of implementation of a B2MML interface has been demonstrated by several WBF members. Typically, a new interface for an application can be created in about 5 weeks instead of efforts of up to a year to build a custom interface between ERP and an MES. As many vendors and systems adopt B2MML, this single, standard interface replaces many custom point-to-point interfaces, reducing the cost of ownership by eliminating development and support costs. By using a standard interface, a company is not locked into a single vendor's product or a single skilled consultant. Systems at either end of the standard interface can be changed at a minimal cost.

Once this standard interface is integrated into a company's IT infrastructure, deployment to new sites becomes a standard service request, rather than a development effort. The accuracy of information in an automated interface reduces the possibility of error inherent in a manual interface. Once automated, information exchange can be more frequent. This can allow more frequent and accurate planning functions to respond quickly to supply chain changes. P&G expects to replace daily back flushing calculations of inventory with frequent direct recording of production and consumption. Just-In-Time (JIT) raw material inventory replenishment and urgent production order changes can now be deployed within a very small time window.

The reduced integration cost can also bring supply chain integration to small operations, such as emerging markets, where it has been more cost effective to manually exchange data between ERP and MES systems. The simple XML technology of B2MML should make it more cost effective to automate interfaces to systems as simple as a spreadsheet. The existing file transfer infrastructure could be used at no additional cost. Even better, with no interface development cost, more sophisticated MES systems may now be cost effective. Manual error elimination, rapid supply chain information flow, and cost reductions could be implemented even in these already low-cost and low-volume operations.

Vision for ISA-88 and ISA-95

Our vision for an integrated batch making information system has long been clear. Today we are taking the last few steps to make it a reality. The diagram of this vision (Fig. 13.1) shows how we intend to use our key MES system and our ERP system to build this integrated batch making information environment. Our batch execution systems vary greatly, from simple Programmable Logic Controller (PLC) programs, to sophisticated batch control systems. Our key MES system will soon be able to store and integrate batch history from all sources in a single ISA-88 structure.

A breakdown of the four batch history methods is illustrated in Figure 13.1. A single Batch Interface Table becomes a point of integration for all batch history

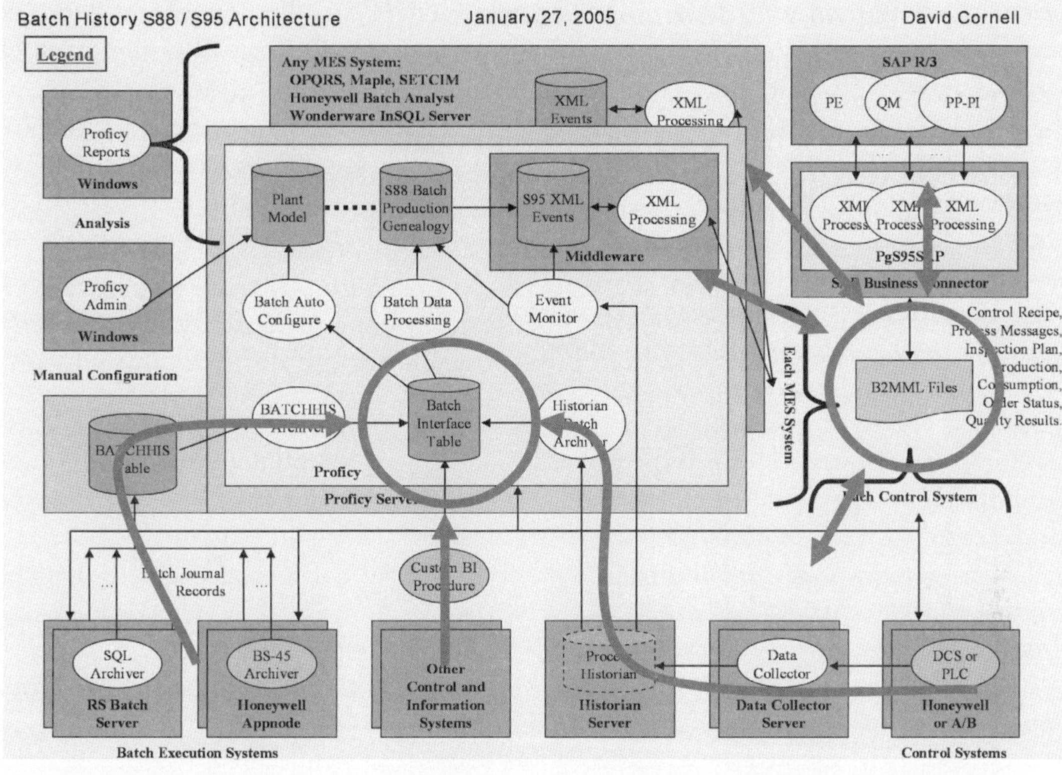

Figure 13.1. Circles highlight the points of integration, batch history, and ISA-95 B2MML.

from batch control systems. Some batch control systems archive their data to a common relational table. A module reads this table into the Batch Interface Table. There is also a module to extract batch history data from specifically named tags in a historian into the Batch Interface Table. These historian tags are populated by PLC code and act as a single entry transaction buffer for batch history data. Other systems can insert their own data into this Batch Interface Table using SQL. The event detection models could also be configured to derive batch data from event triggers and data in a historian if no batch data structure can be constructed in the control system.

The single point of integration to ERP is the B2MML interface. Two of our MES systems use the same B2MML messages. Interfaces may also be developed for control systems, either by their vendors or us, to read B2MML information directly when there is no need to pass it through an MES system. The schedule for production, for example, could go directly to an intelligent control system, as well as the MES system.

"P&G ISA-95" as a B2MML Application for ERP

To begin our B2MML implementation with ERP, P&G created an application called P&G ISA-95 to translate between ERP messages and B2MML. We chose the mapping tools that already had an infrastructure with a support organization and development resources available. We will develop only the minimum functionality required to meet our short-term needs with this application. We expect to phase this out in favor of the ERP vendor's product as soon as possible (Fig. 13.2).

The transfer of B2MML messages between the single XML message server in each continental region and the Manufacturing Systems is done by File Transfer Protocol (FTP). The sending system is responsible for handling communication

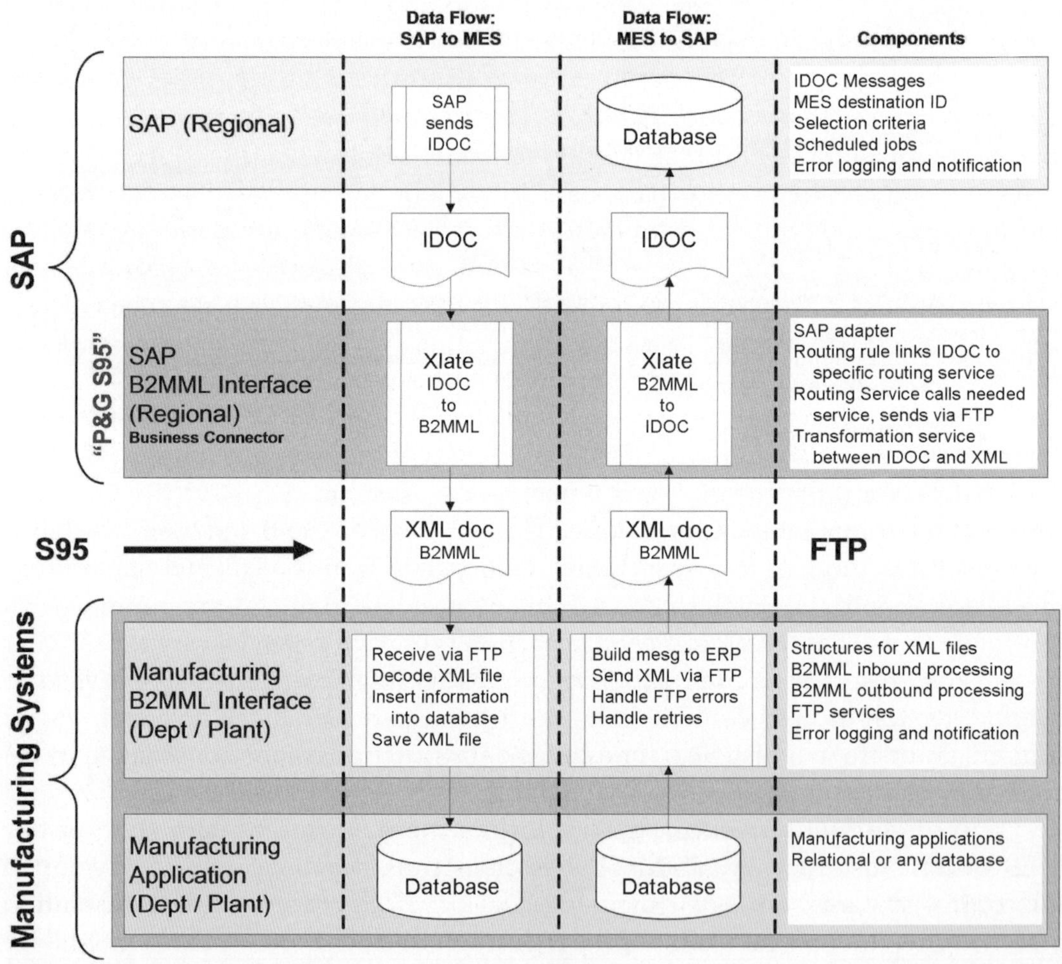

Figure 13.2. ISA-95 B2MML between ERP and MES.

errors and retry attempts. This method was the easiest to implement for our pilots. We are investigating other methods, including a publish-subscribe mechanism already in place at P&G.

P&G Integration Projects

Last year, we completed a pilot in our Family Care business with MES receiving Production Schedule information from B2MML, sent from ERP. This schedule download function is being deployed to many more of our businesses.

Last year, we also completed a pilot in our Fabric and Home Care business with their internally developed MES application. It implements automated supply chain integration by receiving the schedule with the bills of materials and sending material production, material consumption, and order confirmation. This application is being deployed around the world in this business this year.

This same functionality will be implemented in our key MES system and piloted in our Health Care and Beauty Care businesses this year. We are adding the ability to send Quality information related to the produced lot of material for use in the lot release decision in ERP. The Health Care business will also be prototyping two of the Batch History methods in MES. Once installed, we will have the first implementation of our vision of ISA-88 and ISA-95 integration from the plant floor to ERP through MES, using a close approximation of B2MML rather than custom interfaces.

The Future

The Future of ISA-95 at P&G

Our next step will be to migrate from our interim solution for B2MML to those supported and delivered as products from ERP and MES vendors. Our interim solution will be enhanced only when the products from our vendors either do not have such a function or when we are unable to upgrade our vendor products due to other factors.

We will encourage the owners of all MES systems in our company to migrate to our key MES system. It will then become the primary point of integration among our Manufacturing Operations, maximizing the efficiency of information exchange in our Consumer-Driven Supply Network.

If this migration is not possible, we will require the owners of MES systems to use some type of B2MML to exchange information with ERP. We want deployment, reconfiguration, and support of our ISA-95 integration to be a simple service request. The universal use of B2MML will allow this.

I will represent P&G at the industry interoperability workshops. I will also be the key contact with our ERP vendor on their B2MML development. I will also continue my work representing P&G in the WBF working groups for B2MML. It is yet to be determined if I will represent P&G on the ISA-95 committee; P&G is currently not represented.

The Future of ISA-88 at P&G

A common vendor-supported and standard solution for batch history in P&G has been a goal for much longer than a standard ISA-95 implementation. When our key MES vendor fully develops its ISA-88 model and we deploy it around our company, that vision can be realized. However, this ISA-88 integration, while valuable on its own for common reporting and analysis, is only a small part of the need for integration.

Our businesses wish to respond almost in real time to changes in the supply chain. Therefore, our planners will be scheduling each batch of production individually. Both ISA-95 and ISA-88 integration and standards are required if we are to realize this vision in a cost-effective and efficient manner. MES will provide a common ISA-88 batch history repository across our batch making systems. B2MML that is implemented and vendor supported in MES and ERP will bring ISA-95 integration to the batch processing area. P&G is currently represented on the ISA-88 committee by Dave Chappell. He may also participate in the industry interoperability workshops.

The Future of ISA-95 in the Industry

The future of ISA-95 across the industry does indeed look bright, yet there is still a lot of work to do. The efforts to improve the ISA-95 standard by ISA and the effort to improve B2MML by the WBF must continue. Without a complete definition for integration, vendors must continue to develop custom interfaces.

The Process To Business (P2B) and Discrete To Business (D2B) Interoperability Working Groups are an excellent forum for the promotion and deployment of ISA-95 and B2MML. We should encourage all our vendors to participate or at least subscribe to the reports from this effort. These working groups started as a forum for our ERP vendors to solicit input for their ISA-95 development. They are open to include all ERP, MES, and manufacturing vendors and customers.

CHAPTER 14

How to Implement an MES without Making a Mess

Presented at the WBF
Make2Profit Conference,
May 24–26, 2010, by

Gerald (Jerry) Sandoval
MES and Process Automation Consultant
gasandoval@gmail.com
G A Sandoval Consulting, LLC
5019 Cortland Court
Midland, MI 48642, USA

Abstract

Manufacturing Execution System (MES) implementation projects are difficult, with many challenges that must be addressed in order to not only successfully implement the MES, but also to sustain the MES, once implemented. This chapter addresses the *dos* and *don'ts* of implementing an MES, based on lessons learned from over one hundred MES implementations. The chapter will also propose approaches that will allow an MES to weather the test of time and the inevitable changes associated with it.

Scope

This chapter is applicable to all types of plants that may want to implement an MES. This includes batch, semicontinuous, continuous, discrete, and hybrid manufacturing processes. This chapter is also based primarily on experiences of implementing MESs where the business and control systems are already in place.

This chapter does not specifically address the issues associated with implementing MES in a new plant. It is very desirable to design a new plant from the ground up with MES in mind. Although all the topics covered in this chapter are applicable to new plants, there are additional challenges when implementing an MES for a new plant or for a major expansion of a plant that will not be addressed in this chapter.

Introduction

To successfully implement an MES, you must understand and address the many facets of the MES project and the MES life cycle. These many facets include the following:

- Project management
- Multidiscipline teams
- Effective communications
- Education and training
- Change management
- Work processes
- Personal interactions
- Systems integration
- Technology

Yea or Nay?

The very first question to answer when considering implementing an MES is whether or not implementing an MES provides value. What are the costs versus benefits? Is the total cost of ownership understood? The cost to implement is only part of the long-term cost of ownership. The costs to operate and maintain the MES have to also be factored in. The following are some potential value propositions to consider:

- Will it reduce costs (i.e., reduce manpower, waste, inventory levels, and create tangible savings)?

- Will it improve capacity (i.e., reduce time to produce and reduce cycle time)?

- Will it improve quality (i.e., reduce or eliminate off-spec product)?

- Will it improve time to market (i.e., introduce new products more quickly)?

- Is it an enabler (i.e., enables Lean manufacturing or some other value-providing system)?

The bottom line is, don't implement an MES just because everyone else is! Implement it because MES provides value to your plant.

Define and Manage Expectations

Two issues that plague many MES implementation projects are (1) not having a clearly defined functional scope and (2) obtaining sign-offs from the sponsors of the project. It is extremely important that everyone is on the same page as to what the project will deliver. It is also important to not only define what is within scope, but also to clearly state what major functionality is not within scope, in order to properly set expectations for all involved.

Be prepared to deal with the "oh, you can do that too" revelations during the project. This happens frequently as project team members from the plant or business better understand the capabilities of the tools and systems being used and the potential value that they may provide. These situations can lead to scope creep if not carefully managed. Put in place a scope change process and diligently follow it. Finally, only accept scope changes with a signed agreement that clearly states ramifications such as increased cost and new delivery dates. Remember, clear communications help manage expectations.

Get the Buy-in from *All* Involved

In order for an MES implementation to be successful over the long term, the plant must *own* their MES. If a plant does not own their MES, it will fail from neglect to effectively address the necessary operational maintenance. The following are steps that should be taken to help foster ownership of an MES:

1. Identify a champion for the MES project, preferably from the plant or business management. Care must be taken to avoid a management edict, where the plant resents the project, rather than owns it.

2. Assign a focal point person to deal with the daily details of the MES. This person should be involved with the MES implementation project in order to gain a thorough understanding of the implemented system.

3. Ensure that the Operations staff is onboard with the project. If the Operations staff is not onboard, then the project is doomed, especially if they are being asked to perform specific actions that are critical to the daily operation of the MES. Ask the question, "What's in it for them?" If the answer is more work, you are going to get push back, and more than likely, missing or inaccurate data if you are asking them to input data. To address this issue, design the MES to "make it easier to do it right than to do it wrong" by automating as much as possible. When asking for input, don't ask for something that you already know.

4. Verify that all organizations involved in implementing an MES accept the long-term support responsibilities. Proper support is mandatory for long-term success of an MES.

Team Work Is Essential

Because of the complexity of MES, it is rare that one person can implement and support all the component systems that make up an MES. Thus teamwork is essential to implement and support an MES throughout its life cycle.

In order to be successful, start by identifying all organizations involved in implementing an MES. This includes organizations that support business systems, manufacturing systems, control systems, Engineering, Operations, IT, and so on, as well as representatives from key vendors. Put together a multidiscipline team, drawing resources from each organization, and agree on the responsibilities of each organization. Because an MES routinely involves integrating and interfacing multiple component systems, it is very important to agree on the owner and decision maker for each component system. It is also necessary to agree on the details of the design of each interface between these systems. This teamwork is critical for successful implementation and operation of an MES.

Project Team Commitment

Commitment to success is *mandatory* for all project team members. Project team members with multiple projects or other responsibilities can, and most likely will, negatively impact the project schedule. Project team members with responsibilities to keep the plant operating can become a problem because of conflicting priorities. In most cases, team members with plant responsibilities are critical to the success of the project because of their in-depth knowledge of the plant. Project management must work with plant management to clarify project priorities for these team members.

Some project team roles need to be full time, and others may be on a temporary basis. Temporary team members must be responsive and deliver assignments on time. Carefully manage temporary team members, and avoid placing them on the critical path of the project plan. If they do end up on the critical path, then project management must address the situation as quickly as possible to avoid project delays. The bottom line is that a lack of team commitment can be disastrous!

Educate and Communicate

Appropriately educate *all* involved about what MES is and what the MES project will provide to the plant. Communicate what is in scope as well as what is *not* in scope. Communicating what is *not* in scope can reduce scope creep and eliminate surprises at the end of the project (e.g., when the project exceeds time and budget).

Plant management must understand and approve the scope, benefits, and cost of the MES project. Project team members must understand their responsibilities and deliverables in order deliver an MES that meets the functional requirements on time and within budget. Plant operations must understand and accept what the MES will provide for them and how it will impact their work processes. Involved organizations must accept their responsibility in implementing and supporting the MES.

Be prepared to educate new team members as they join the project team, not only during the MES implementation project, but also throughout the life cycle of the MES from beginning to end. Communicate throughout the project, and address issues as they come up.

MES Design Approach

The MES design does not involve the design of the software—it is the design of how to implement MES functionality that effectively maps to the specifics of the manufacturing facility and the manufacturing processes in use.

The basic rule of thumb is to develop a simple yet robust design approach by using the "Keep It Simple, Stupid" (KISS) principle. The KISS principle states that simplicity should be a key goal in design and that unnecessary complexity should be avoided. So start out with the basic requirements. However, you should develop a long-term master plan to ensure that the initial implementation is compatible with the long-term goals. You can always expand functionality later—if the team members can be found.

Things change, so your MES design should plan to accommodate changes. For example, new products, new raw material, control program changes, new process cells or units, and so on routinely change over time. Be careful not to hard code references to things that might change, no matter how remote that may seem at the time.

Leverage industry standards like ISA-95, ISA-88, and Business To Manufacturing Markup Language (B2MML). ISA-95 defines Manufacturing Operations Management (MOM) and the Production Operations Management (POM) subset that most closely aligns with and defines the functionality of the MES. ISA-88 predominately defines batch control standards but also defines recipe management functionality across all levels. B2MML is a set of Extensible Markup Language (XML) schema that standardizes data and transactions to interface business to manufacturing systems. Following industry standards makes it easier to integrate modules from multiple vendors.

Understand the Current Systems

In order to develop a good MES design, you must first understand the current systems and associated processes.

First Steps

The best way to understand the current system is to start out by taking a tour of the plant manufacturing facilities to gain a firsthand understanding of how the plant operates. The tour guide usually provides insights into issues and potential opportunities for improvements that you might not otherwise get. The plant tour also allows you to more easily visualize the areas, process cells, and units, and it also familiarizes you with the equipment naming and terminology used by the plant. The next step is to interview individuals with key roles in the plant to understand how the plant operates and what each role contributes to the operation of the plant. Because of the plant tour, you will become more familiar with the plant design and the terminology used by the plant. As a result, you should be able to communicate much more effectively. Again, these interviews will provide opportunities to gain insight into issues and potential opportunities for improvements.

Manufacturing Processes

Understanding the manufacturing processes is the foundation for implementing an effective MES. Diagram the manufacturing process down to the unit level.

Map the material flows through each area, process cell, and unit. Determine which areas, process cells, and units are in scope or may provide challenges.

Control and Data Acquisition Systems

Control and data acquisition systems will be your primary source of automated data, and essentially, your window into the manufacturing process. Identify all relevant control systems and capability, and identify all data acquisition systems. Analyze whether or not all required data and events can be captured at the desired level of accuracy.

Manufacturing Systems

Identify manufacturing systems that are currently in use. Determine the level of integration required. Does the level of integration require tightly coupled or loosely coupled integration? Determine data flows between systems and whether or not interfaces exist or need to be put in place.

Business Systems

How is the plant using its business systems today? Diagram how each business system defines and maps to the actual manufacturing process at the area, process cell, and unit levels. Also understand how the systems define storage locations, and understand the difference in the way that the systems define the manufacturing process. For example, the systems may only be concerned with the process cells and not the units within the process cells. Determine which data and processes are expected to be handled by MES. Use your research to design the MES to effectively integrate *all* the component systems to work as one system.

Understand Production Work Processes

Understanding the production work processes is critical because this is how the Operations staff interacts with the manufacturing processes. First, understand how the production work processes are supposed to work. Then, understand how they *actually* work! Differences are inevitable! Right or wrong, people develop shortcuts and simplify to make their jobs easier. Understand the exceptions!

Remember, Operations staff will do whatever they can to keep a plant running. Also know that they will be reluctant to admit to what they do in the dead of night to get their job done. Be prepared to deal with the exceptions or eliminate them! Use your research to design the MES to effectively support and automate the production work processes.

Understand Constraints

When implementing an MES, you must understand not only the manufacturing processes and work processes but also which component systems or vendors must be used and which ones offer you some flexibility. This understanding will help narrow down the choices about which products can be used to address the gaps in the MES design. Remember that MES components are almost always multivendor and that you usually can achieve better integration using fewer vendors that have component systems that integrate together well. The following are questions that need to be answered:

- Does corporate headquarters determine which systems are used, or do they allow some latitude at the plant site level (i.e., centralized versus decentralized management)?
- Is the business system that is used a corporate mandate?
- Are you obligated to use the control systems that are already in place?
- Are specific vendors preferred in order to leverage existing contracts with favorable discounts?

Man versus Machine Interaction

Automate as many transactions as possible, which should reduce or eliminate the human factor (i.e., human errors). Automation should also improve data accuracy and provide more timely data. Instrumentation can be read via control and data acquisition systems and use Radio-Frequency Identification (RFID) or barcode mechanisms, where appropriate.

Understand the interaction of manual transactions with automated transactions. For example, can automated transactions fail if manual transactions are not performed in a timely fashion? Minimize or eliminate automated transactions that require manual intervention for completion. For example, material movement transactions that require manual input of material lot number or storage location before posting should utilize mechanisms such as barcodes or RFIDs to provide the material lot number or the storage location. This can be done in place of requiring a manual input.

Understand the Ramifications of ERP

Most ERP systems are capable of handling just about any manufacturing situation. However, the more that you ask an ERP system to do for you, the more transactions and data that the ERP system asks from you! Data collection and entry can increase significantly, if not exponentially, depending on the complexity of the manufacturing process. For example, entering the material ID and quantity consumed is much easier than having to enter this information along with additional items such as the storage location and material lot numbers. If the system asks for the storage location and the material lot numbers, then it can cause the need for many more transactions because each time these items change, the data need to be captured and more transactions need to be posted. Some businesses, such as pharmaceuticals, may need to capture detailed data for regulatory reasons, and others may not. Determine this requirement early in the project. If the detail is not needed, then the storage location and material lot number can be set to constant default values for each raw material, which will eliminate the need for multiple transactions and greatly simplify the data collection required. If detailed transactions are required, and you cannot automate the data collection and data entry, then can you afford an army of people to do it manually? Remember that although it may be possible to do something in an ERP system, it may not be practical, so keep it simple.

Avoid the Tower of Babel

MES inherently involves multiple systems, each with its own databases and data code systems. Therefore, data code synchronization is mandatory across component systems if an MES is to work at all. For example, each component system must have common code definitions for material IDs, storage location IDs, units of measure, process cells, units, and so on. Without common code definitions, the multiple component systems cannot effectively share or exchange data.

In order to address this issue, a robust data code management mechanism is essential. Ideally, only one database would be required, but this is not realistic when multiple systems, usually from multiple vendors, are involved. So the data code management mechanism is a formal procedure to ensure that all data codes are updated in a coordinated fashion by manual or automated methods. Whichever approach is used, one database must be considered the master database for each type of data, and the other databases must be synchronized to it. Individuals working with the component systems that make up the MES must not circumvent the data code management mechanism, because extraneous uncoordinated changes will lead to the component systems being out of sync with each other. This will cause failures.

Another issue that must be addressed is the data disparities between component systems. For example, an ERP system may schedule production orders by process cell, but the control system may manage recipes and collect data by unit. The MES must be designed to be able to handle these disparities transparently.

One area that can be a challenge is determining how finished product batch or lot numbers are assigned. This may initially seem easy, but there are situations that can make this very challenging because of the various limitations of interfacing with control systems. Rules must be developed for how, when, and where a batch or lot number is created and shared between systems. These rules must guarantee that no duplicates are created and that all systems that need to use it can access it when required. For example, a laboratory information system needs to have the ability to associate a batch or lot number to an in-process sample where the batch or lot number was assigned within the control system.

Process Control and Data Acquisition

Depending on the complexity of the manufacturing process, Process Control Systems (PCS) may be critical to the automation of a successful MES implementation. The more complex a manufacturing process is, the more important it is to automate data collection, thus delivering accurate data in a timely manner.

In most situations, be prepared to make changes to the PCS. This is because most PCS are designed to control the process and not necessarily capture data relevant for MES. A PCS must have the correct recipe to make a specific product, but it may not have the actual use data easily available or be able to associate these data to the correct work order. A PCS should have accurate instrument readings for quantities consumed and produced, but it may not be able to report material IDs, storage locations, lot numbers, and so on.

Beware

Uncoordinated changes to process control programs can cause serious problems in an operating MES. An innocent change to a process control program can result in missing or inaccurate data in an MES.

Data acquisition systems may be your main conduit into the PCS for accessing and recording data. Each data acquisition system has its own limitations and idiosyncrasies that must be dealt with. One major issue that must be handled is that data must be collected at the correct frequency to achieve the desired accuracy. For example, collecting data once a minute could be a problem if the final quantity consumed is recorded in the PCS but is reset to zero before the data acquisition system captures it. Another issue that needs to be addressed is knowing when to collect data that represent the final data for a particular transaction and associating it to a particular order. The techniques used will vary, depending on the data acquisition system used and the type of manufacturing process. For instance, a continuous process will collect data for a specific time frame, whereas a batch process will collect data for a specific batch.

Test, Test, and Test Some More

Testing is extremely important to the success of an MES implementation. The following lists a series of tests that must be performed to ensure a successful implementation:

- Test all transaction types
- Test data collection from all process cells and units
- Test all routings
- Test for proper handling of all exception conditions
- Validate that all data are accurately collected and posted
- Test all system interfaces, particularly for error handling

Noninvasive, in situ testing is ideal but not always possible. This testing approach tests the real world rather than a simulation and uncovers data collection accuracy and timing issues.

Develop Updated Work Processes

Implementing an MES will result in changes to the work processes. Update the work processes to reflect how to use the MES in normal operation, and more importantly, document how to use the MES when everything goes wrong. Do not let an MES failure be the reason that the plant shuts down or that product cannot be shipped. Design the MES for uninterrupted plant operation. Provide procedures to manually operate business and ERP systems to get product out the door. Provide procedures to recover data from an MES or component system outage. The updated work processes must address multiple scenarios: ERP outage, PCS outage, data acquisition outage, network failures, and so on. Develop a comprehensive change management plan. Change happens, so be prepared to deal with it!

Final Observations

The points made in this chapter only begin to scratch the surface of the many challenging facets involved in implementing an MES. Technology is not necessarily the major issue when implementing an MES. However, technology can be extremely challenging if the selected MES software is not a good fit for the situation, is still under development (avoid version 1), or has requirements that cannot fully be met. Remember, MES implementation projects are, by default, systems integration projects, because it is rare that any two plants have the same combination of business, manufacturing, and control systems. This necessitates multiple systems from multiple vendors.

Good project management combined with an excellent understanding of the technology and the plant is key to a successful MES implementation. But remember, successful implementation is only the first step. Strong ownership by the plant, along with an active management style on the day-to-day operation of the MES, are keys to the successful long-term operation of an MES. Successful long-term operation is the goal. However, the planned obsolescence strategies of today's software systems are a constant challenge to the long-term success of any MES.

CHAPTER 15

How to Use ISA-95 Part 3 for MES Functional User Requirement Specifications

Presented at the WBF
North American Conference,
May 15–18, 2005, by

Jean Vieille
Consultant
jean.vieille@isa-france.org
Psynapses Network
10, rue du Stade
Dompierre-les-Tilleuls, 25560, France

Abstract

Shop-floor activity execution control encompasses many management function-alities, addressing several types of manufacturing operations, from the receipt of raw material to the shipping of finished goods and from production itself to equipment maintenance to inventory movements and material quality tests. These activities fall under different responsibilities, though they must operate collabora-tively under the business management directions. When it comes to specifying such a Manufacturing Execution System (MES) system, creating the User Require-ment Specification (URS) may appear to be a difficult task, leading to confusing and complex documentation and eventually project failure.

Since the ISA-95 standard's initial scope only covered Business To Manu-facturing (B2M) information flows between production execution and business systems, the ISA95 committee needed to pay attention to the execution layer func-tional duties. The first valuable contribution to sort the chaos within this large scope of functional requirements was the MESA model of eleven core functions.

Part 3 of the ISA-95 standard defines a more elaborate functional model particularly suitable for URS development. This chapter describes a formal method for defining MES URS and system adequacy analysis based on ISA-95. This method was used for very different projects such as defining a global MES template for a major tire manufacturer, implementing a fully integrated control system to MES for an animal food company, and specifying MES URS for a paper facility.

Introduction

What's Wrong with MES?

When it comes to dealing with MES, no one has the same feeling about it. The manufacturing viewpoint considers Standard Operating Procedure (SOP) enforcement and data collection to be the core MES duty, while the business viewpoint focuses on managing inventory and fulfillment using data from the shop floor. It is this kind of perception divergence that needs to be clarified. The differences come from two facts:

1. Manufacturing involves both processing and execution management, which are very different perspectives on the same subject.

2. Two communities are involved in manufacturing. First, there is the IT/business managing community, which hardly understands actual manufacturing control constraints and needs. Secondly, there is the operational executing community, which typically doesn't comprehend the business mentality and vision.

From this situation, functional responsibility questions and integration concerns appear among both the systems and people involved in manufacturing business and execution management. This makes structuring requirements difficult and results in extremely challenging assessment processes that often require a unique mixture of skill sets and years of study and practice. ISA-95 and MESA exist precisely to help the emergence of common understanding and best-practices sharing.

What's Wrong with URS?

The first step to build a system is to define the requirements that it will address. While these requirements are sometimes expressed in a very broad and blurry way, a precise definition of the actual needs is a key factor for succeeding in implementing a somewhat useful and productive system.

As an example, the pharmaceutical industry naturally places URS at the very top of the Good Automated Manufacturing Practice (GAMP) cycle (Fig. 15.1). However, Figure 15.1 shows an odd multilink between both URS (under customer responsibility) and Functional Specification (FS; under vendor or integrator responsibility) to system acceptance testing. The FS is a translation of the URS in terms that the vendor's engineering staff can understand. Ideally, the customer can understand it, too. Project quality management has to guarantee URS coverage through a URS and FS coverage matrix mapping URS topics to FS modules.

Further explanation of the V-model may be found in Chapter 17 of this book. URS are not qualified by themselves but are indirectly qualified by agreeing on the FS solution addressing each individual requirement. Other project management approaches have similar concerns and seek to create consistent URSs.

Advanced process control URS methodologies can make extensive use of ISA-88 to ensure the consistency between URS and FS by relying on a common robust functional framework. FS becomes a more elaborated URS, instead of being a brand new solution-oriented thinking extrapolated from a hardly maintained one-shot verbose literature. ISA-95 part 3 plays the same role for MES, allowing a more straightforward mapping between URS and FS.

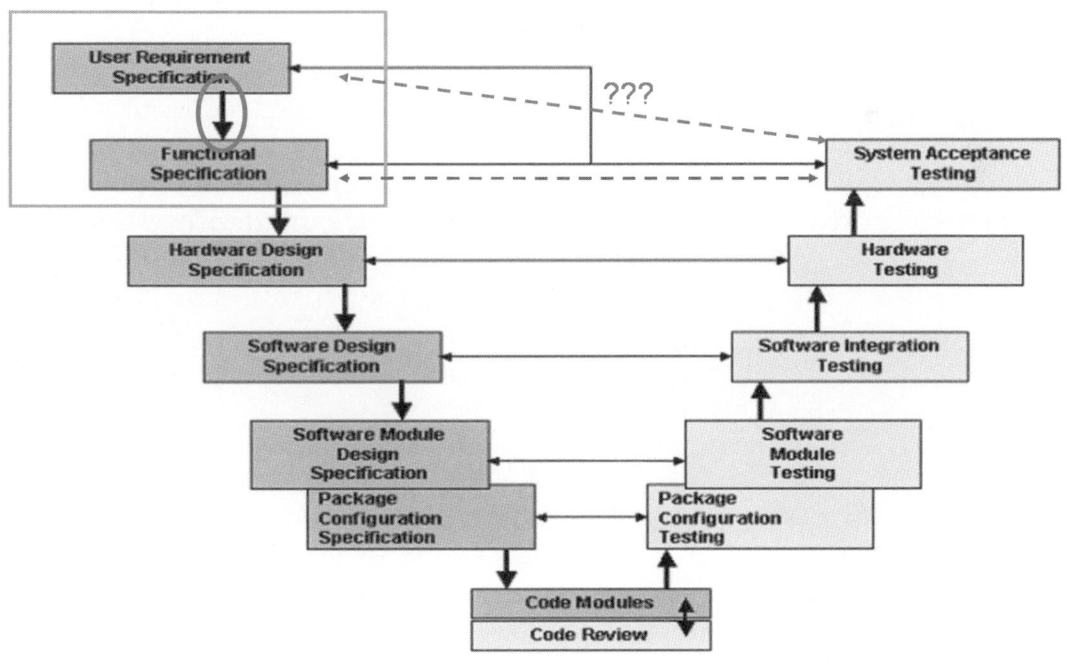

Figure 15.1. GAMP V-model.

ISA-95 Part 3 Functional Framework

ISA-95 provides extensive commonsense models, explanation, and terminology to discuss and address manufacturing execution management in a more consistent and consensual way than ever before. While there are still discrepancies in the standard, a consistent informational and functional framework takes shape.

The part 3 activity management model can be applied to different Manufacturing Operation Categories (MOCs), such as those appearing in the Purdue Reference Model (PRM) shown in Figure 15.2. The standard defines Production, Quality, Maintenance, and Inventory as the four parts of Manufacturing Operations Management (MOM) shown in Chapter 1 (Fig. 1.5). The standard also defines supporting functions that specify common infrastructure requirements (Fig. 15.3). The result is a tridimensional functional framework, comprising MOCs, Core Activities, and Supporting Activities that can be used for URS as well as solution mapping.

Functional Hierarchy

The proposed method for URS development is "process based," as shown in Figure 15.4. Between business planning and work execution processes and management responsibilities, MES concepts are about inserting an execution management layer between business-oriented enterprise planning and execution-oriented enterprise manufacturing. Thus business processes are split or shared between business and execution responsibility domains.

Methodology Overview

The methodology starts from defining actual MOCs as execution responsibility subdomains to drive requirement data collection. Beside this decisive start of the

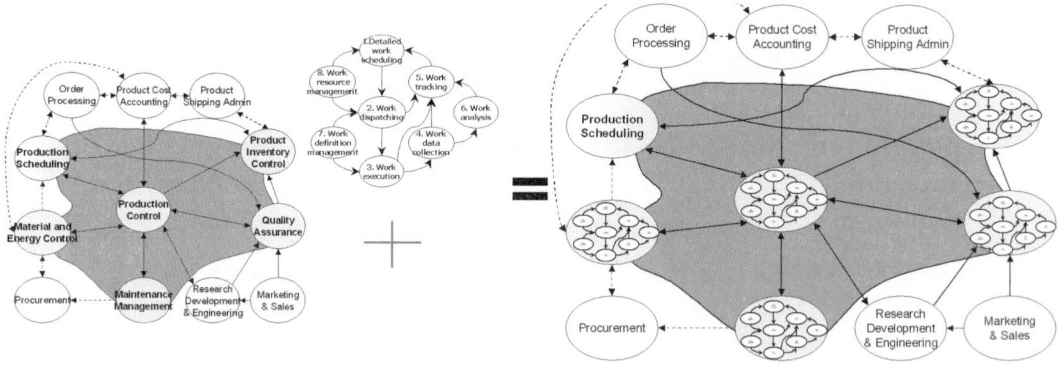

Figure 15.2. Combined PRM and MOC.

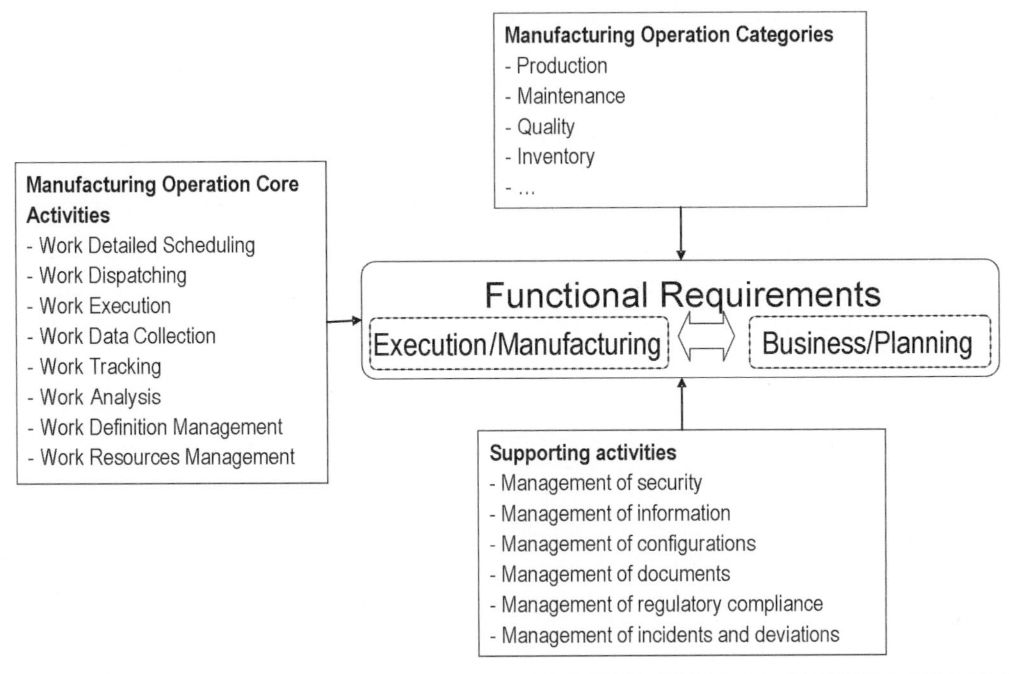

Figure 15.3. Common infrastructure requirements.

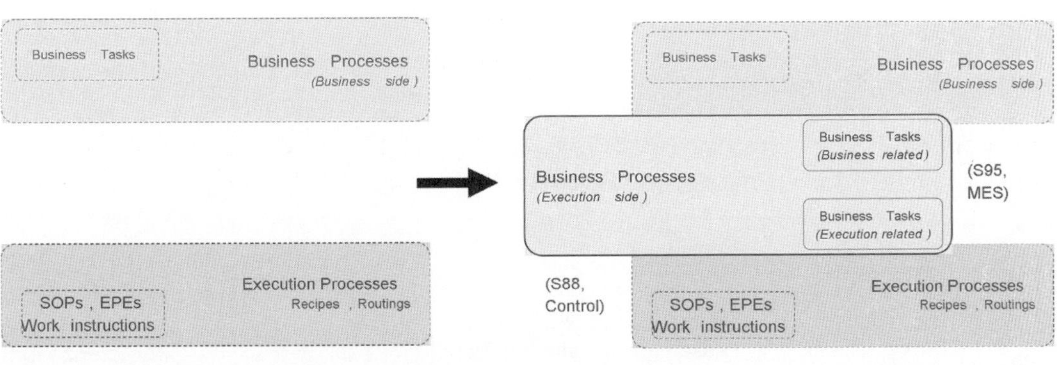

Figure 15.4. MES as a function between business and execution.

organizational and process-based functional architecture design, resource modeling gives substance to the project and support for specific requirements by setting up the current facility capabilities. Subsequently, it defines business processes and identifies the tasks they drive. Business processes define the situations and scenarios that need to be involved in order for a system to achieve its objectives.

Areas of collaboration between MOCs in the execution responsibility domain and the business responsibility domain are highlighted in Figure 15.5, providing a comprehensive example for likely integration requirements. Tasks are characterized, classified, and consolidated to build up or to comply with an enterprise MES functional core system.

Finally, the actual solution capabilities are compared to the functional framework to conduct adequacy assessment and URS-to-FS mapping. Figure 15.6 shows the URS development sequence.

Methodology Hints

Phase 1: Technical Modeling

Technical modeling describes the target facility. It includes five steps to define MOCs and resources, both MOC specific and shared (Fig. 15.7). Resources can be defined independently, while working segments (i.e., a generalization of process segments) are built from these resources. This initial step is mainly about the system configuration that can dynamically change over time, rather than the system design itself.

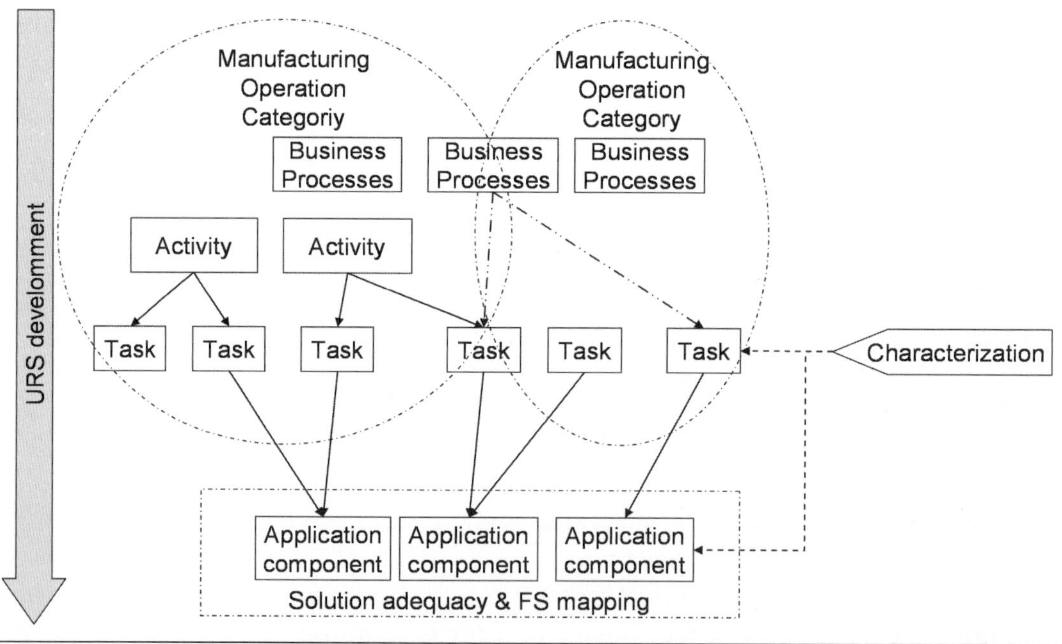

Figure 15.5. URS development from MOCs.

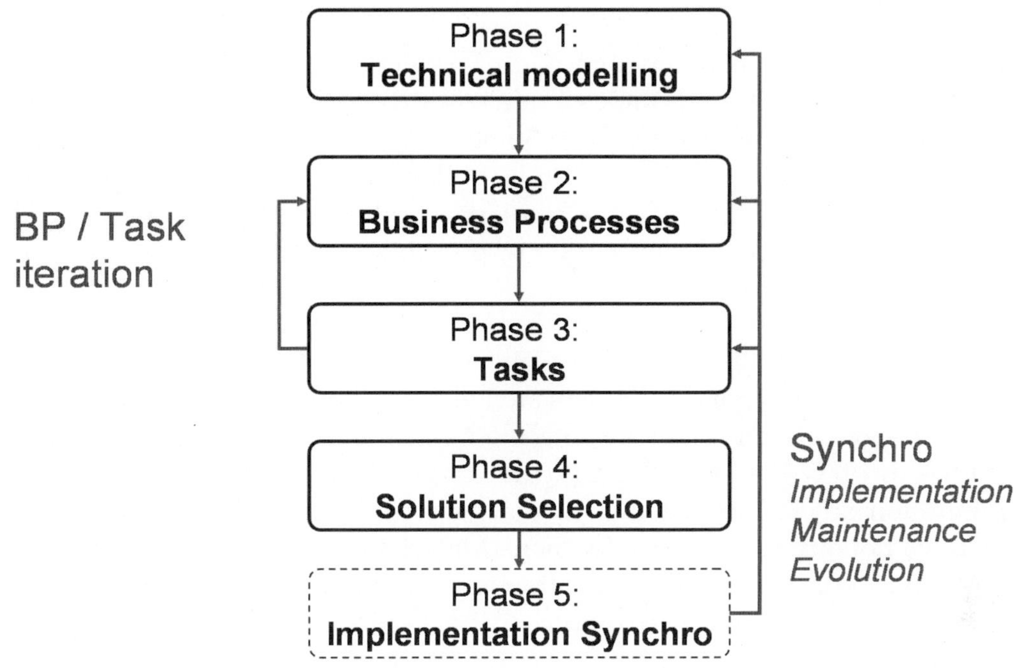

Figure 15.6. URS development sequence.

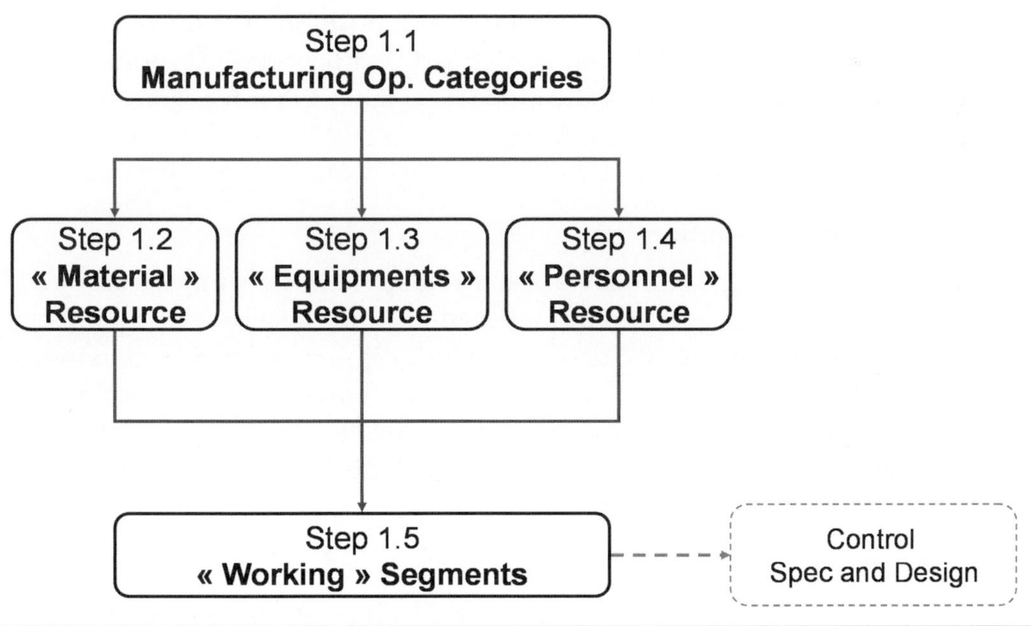

Figure 15.7. Technical model of a target facility.

Step 1.1: MOC

ISA-95 part 3 defines the following MOCs: Production, Quality, Maintenance, and Inventory. Other or different MOCs can also be defined (e.g., outbound logistics, internal transfers, tooling, cleaning). Sometimes several MOCs can be defined for a given activity domain. For example, bulk production and packaging can be defined as specific MOCs instead of being part of the same Production MOC. The key words behind MOCs are execution responsibility and typology.

Step 1.2: Material Resources

At the URS level, only material classes are identified. Material classification allows specifying segment inputs and outputs, filter information, associated specific functionalities or processes, and so on. Actual materials are initialized at the system configuration and will be dynamically managed by the user. The expected number of material definitions can be stated here. Material properties can be defined at a further development stage. Classification is multidimensional and should be consistent with business modeling options

Step 1.3: Equipment Resources

Equipment can be defined according to the ISA-95 physical model (although this is not mandatory if the enterprise already has one). All main (i.e., schedulable) pieces of equipment are identified, classified, and positioned within the physical hierarchy. Equipment properties can be defined at a further development stage. Equipment identification and classification is the main input for work segmentation and system user access definition.

Step 1.4: Personnel Resources

At the URS level, only personnel classes are identified. Personnel classification allows users to specify segment-required qualification, system access, and so on. Actual persons are initialized at the system configuration and will be dynamically managed by the user. The total average number of individuals can be stated here. Personnel properties can be defined at a further development stage. Classification is multidimensional and should be consistent with business modeling options.

Step 1.5: Working Segment

Working segments define the actual work capabilities of the facility and associate the corresponding resources. Segments are hierarchically defined from a very broad perspective (e.g., the entire factory making a particular brand of products) to the smallest processing capability (e.g., a single ISA-88 phase), although the most useful granularity may correspond to ISA-88 unit procedures. They are part of the general control and automation architecture.

Phase 2: Business Processes

Business processes define the highest functional requirement level. They illustrate situations and task activation scenarios, which can be manual, semi- or fully automated, or hierarchic. In a hierarchic scenario, high-level processes activate lower-level processes, and elementary processes are considered tasks. Some examples of business processes are the following:

- Bulk material receipt by truck or train
- Palletized material receipt
- On-receipt, on-line, or on-shipment analysis
- Bulk product shipping
- Internal bulk transfer
- New order processing
- Order cancellations from execution or business
- Unsolicited order reporting
- Order completion reporting
- Process improvement suggestion processing
- Material loss reporting

Business processes can be compared to manufacturing processes: ISA-88 recipes (i.e., business processes) activating equipment procedure elements (i.e., tasks). An example of this is shown in Figure 15.8. This example makes use of Business Process Modeling Notation (BPMN), a high-level workflow description language that (in the case of computer-controlled processes) can be mapped to design-oriented languages such as Business Process Execution Language (BPEL), BPEL for Web Services (BPEL4WS), and Business Process Modeling Language (BPML).

Business processes can be categorized at will to expedite extensive and structured requirement gathering:

- Execution management processes deal with work organization and execution.

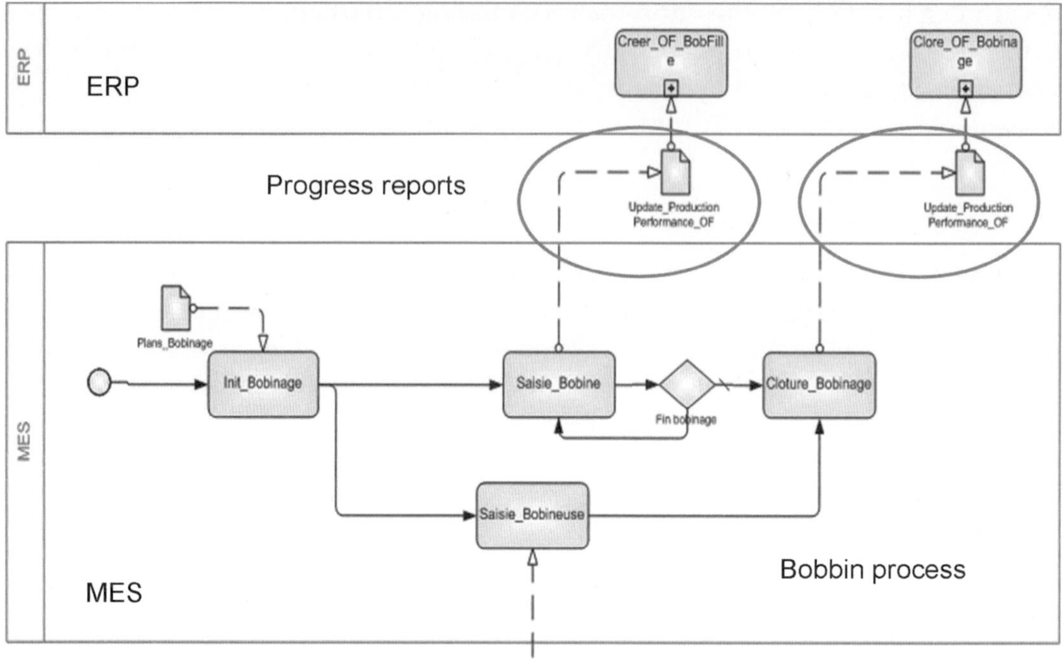

Figure 15.8. Example of an MES recipe.

- Resources management processes deal with resources rather than work execution.

- Operation management processes deal with operation performance, including dashboards, performance indicators, and general activity reports. They are not linked to work orders or resource control.

- Repository synchronization processes aim to maintain consistency between configurations databases in all systems involved in the previously listed processes. The need for such processes largely depends on business management policies and supply chain dynamics.

Business processes can be partially or fully automated, and they can also be manually controlled. Business processes can be MOC specific, generic for several MOCs, or shared by several MOCS. For example, an on-line Quality Control (QC) test's process can involve tasks within Production and Quality MOCs.

Business processes can be confined within business systems, within execution systems, or they can cross the business and execution system boundaries. Thus business processes are the primary input for identifying interfaces and corresponding messages and transactions.

Phase 3: Tasks

Step 3.1: Task Definition Overview

The business process definition phase highlights the tasks involved in their execution. A list of tasks can be established as the basis for functional definition. These tasks are organized following the ISA-95 part 3 activity model for each MOC (Fig. 15.9).

The next steps will elaborate, refine, consolidate, and classify this task list. Task consolidation attempts to reconcile similar tasks into fewer generic task classes. From the optimized task list, business processes are amended to match a more global functional modular definition.

Task definition and business process alignment are executed simultaneously and iteratively until satisfaction. Performing this exercise the first time will result in a consistent set of hopefully reusable functional components. Future projects will make use of these components as much as possible. They will improve and complete them to form the enterprise MES components core system.

Step 3.2: Description

Task description includes two dimensions: (1) characterization that defines usage attributes and (2) justification and functional description that defines the services to be provided. Typical characterization attributes are as follows:

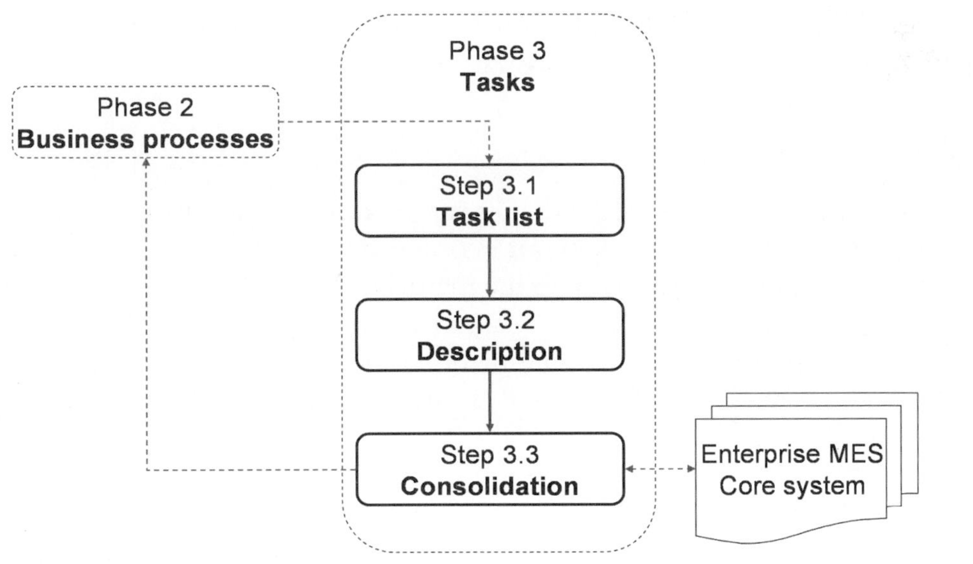

Figure 15.9. Defining tasks.

- Physical level defines the position of the task in the physical hierarchy.

- MOCs restrict the task applicability to particular MOCs.

- Segments restrict the task applicability to particular working segments.

- Technical constraints define the task environment and condition of usage.

- Dependences indicate that a task execution depends on the existence of other tasks. This is used to justify apparently non-value-adding tasks.

- Task style defines the type of functionality the task achieves (e.g., real-time or transactional control, data storage, knowledge management, data analysis, simulation and modeling, etc.).

- Responsibility specifies if the task is under the responsibility of business or execution.

- User specifies the personnel classes who have access to the task.

- Justification gives technical and financial reasons for implementing the task and associated implementation priority.

- Information defines the consumed, produced, manipulated, or presented information based on ISA-95 parts 1 and 2.

Functional description will actually explain the expected task behavior in normal and exceptional operation conditions

Step 3.3: Consolidation

The MOC and process-oriented specification produce a basically inconsistent list of tasks. The last step is to sort them out in order to have a reduced, consistent, reusable list of tasks that hopefully map the maturing enterprise MES component core system (Fig. 15.10). This object-based classification will greatly simplify and harmonize execution processes by propagating best practices and questioning unjustified specifics. For example, similar tasks over multiple MOCs can be as follows:

- Harmonized between MOCs if they are identical but separate (e.g., work instructions and Microsoft Word)

- Kept specific by MOC if they are different (e.g., work definitions and AutoCAD)

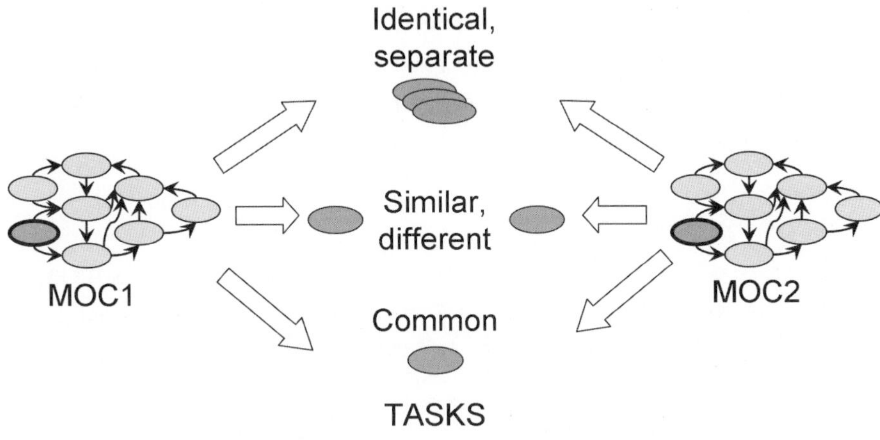

Figure 15.10. Sorting the tasks.

- Shared between several MOCs if the same task instance addresses the functional needs for several MOCs (e.g., managing personnel information)

Phase 4: Solution Selection

Selecting the solution may follow the approach illustrated in Figure 15.11.

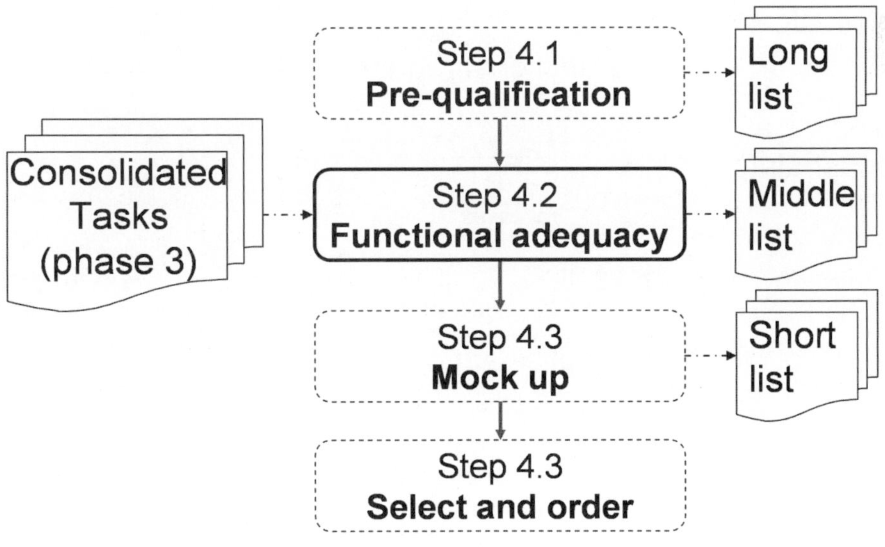

Figure 15.11. From tasks to solution.

Step 4.1: Prequalification

The purpose of this step is to preselect a limited list of the best possible solutions based on nonfunctional criteria (e.g., company and product records, local support, integration capabilities, technology, etc.).

Step 4.2: Functional Adequacy

The purpose of this step is to match the functional requirements (consolidated tasks) with the solutions capabilities (Fig. 15.12). When considering actual solutions against functional requirements, one has to consider the different aspects of the solution:

1. Inherent capabilities that can provide embedded solution functionalities that can map to one or several tasks

2. Integration capabilities that can leverage inherent capabilities to fulfill the requirements (i.e., what can be done within the system to develop interaction between standard modules without impacting the solution integrity)

3. Specific solution developments that are needed to complement the solution inherent capabilities and integration artifacts

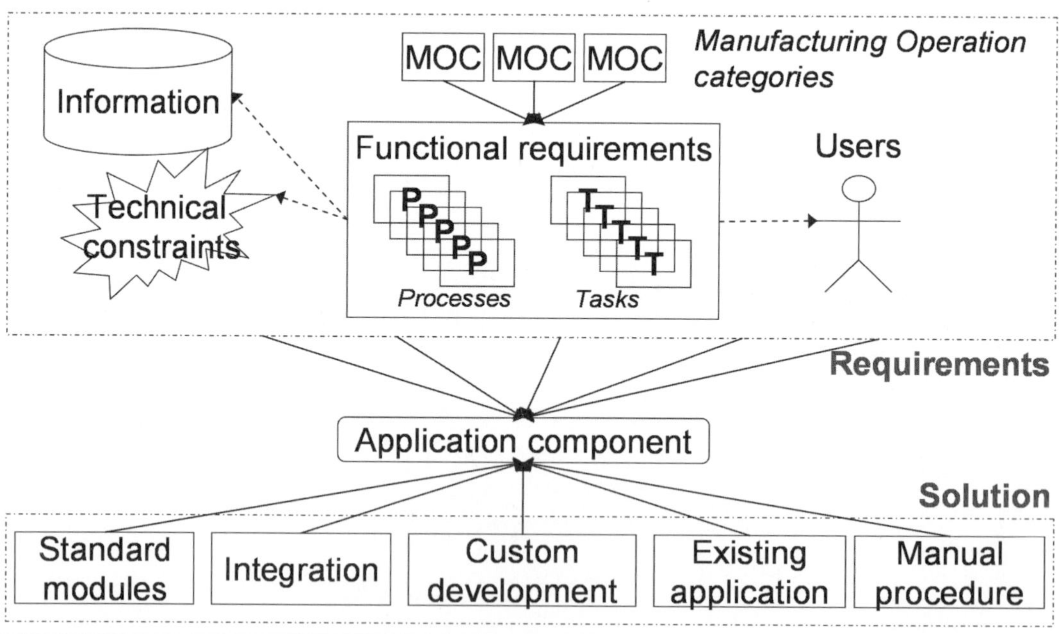

Figure 15.12. Matching consolidated tasks with solutions.

4. Existing applications that can still operate in the target solutions

5. Manual procedures that can be applied instead of a software implementation

The possibility for a particular element of the solution—an application component—to address more than one task requirement is assessed against technical constraints, information consistency, user interaction, and applicable MOC. This is an objective process that can help reduce the number of possible solutions by eliminating any solutions that inherently don't cover enough requirements using aspects 1 and 2.

Step 4.3: Mock Up

This step will implement different scenarios involving a set of tasks and system integration features to test the possible remaining solutions. The result is a shorter list that contains only the suitable solutions in terms of integration and interfacing capabilities and actual feasibility. This will improve the subjective perception of the solution, but it won't assess its behavior in normal operating conditions and scalability.

Step 4.4: Select and Order

This is the final step, where technical considerations cease changing for financial negotiation

Phase 5: Implementation Synchronization

The URS stage is not really worth carrying out when the following is true:

- Only initial system implementation is considered
- URS is forgotten as soon as the first functional specifications are released

However, the following should be taken into consideration:

- The project outcome needs to be controlled
- Enterprise best practices and know-how are considered as valuable assets
- The system life cycle has to survive commissioning

If this is the case, then the following steps should be taken:

- URS should be in sync with actual implementation. Unspoken requirements showing up during the system implementation and commissioning have to be included in the URS.

- Each new or modified requirement should be engineered at the URS level to maintain the consistency with the core system, to address issues from a broader perspective, and to make the corresponding improvements available at the enterprise level.

- The user and solution integrator must agree on the documentation form and be committed to use it as the highest-level contractual interface between all project actors.

Conclusion

MES implies a clear perception of manufacturing control and planning. Conducting successful MES projects and building flexible, evolving solutions is highly challenging. It can only be achieved thanks to a high-level of enterprise maturity and a full commitment to ongoing improvement.

The ISA-95 standard brings a common vision between business and execution. The whole standard is an excellent tutorial for understanding MES functions and involved information from a universally shared perspective. Part 3 is particularly useful for structuring a URS. The process- and MOC-based requirement analyses improve the URS's completeness and transparency. The proposed methodology follows the spirit of the standard by doing the following:

- Encouraging responsible user involvement in the system definition

- Helping mutual understanding between project actors

- Being independent of the technical solution, thus allowing full deployment flexibility and system evolution

Establishing and Maintaining ISA-95 Standards Is Great . . . But It's Even Better if They Are Consistently Executed across All Operations

Presented at the WBF European Conference, November 10–12, 2008, by

Jean-Luc Delcuvellerie
Director
MES Product Management
jean-luc.delcuvellerie@apriso.com
Apriso Corporation
267, Bd Pereire, 75017, Paris, France

Abstract

The activity model defined within the Manufacturing Operations Management (MOM) layer of ISA-95 is much better than prior Manufacturing Execution System (MES) models, since it takes into consideration both production and non-production operations, such as Quality, Inventory, and Maintenance. Because solutions have historically been implemented by different organizations and software providers, operations data and intelligence provided at the execution level still tend to be disconnected. Siloed architecture models are prevalent, including those providing MESs, Quality Management Systems (QMSs), Warehouse Management Systems (WMSs), and Enterprise Asset Management (EAM).

Furthermore, with the accepted reality of globally distributed operations, a MOM framework must address how to manage global operations with a capability of identifying and distributing global best practices. Large companies now seek to implement MOM applications in the same way—as an enterprise application to support their global operations.

New solutions now exist based on manufacturing Business Process Management (BPM) and Service-Oriented Architecture (SOA) to support this requirement. This chapter will illustrate three examples of how an enterprise MOM solution has been successfully implemented:

- From product design in Product Life-cycle Management (PLM) to complex assembly process execution in MOM in Aerospace and Defense (A&D)

- From material purchase in Enterprise Resource Planning (ERP) to components receiving in MOM in Automotive

- From quality planning in ERP to lot inspection in MOM in Consumer Goods

Benefits of the ISA-95 Part 3 Reference Model

A real innovation, compared to the prior MESA MES reference model focused on production-related activities, was introduced in the ISA-95 reference model by explicitly extending MES activity beyond pure Production to all manufacturing operations including Maintenance, Quality, and Inventory, as described in Figure 16.1.

Figure 16.1 shows how this model includes inventory operations, which were not explicit in the prior MESA MES model and not detailed at all (in fact even considered external, as part of logistics) by the recently revisited version MESA model.

The ISA-95 reference model is symmetric, which makes it clear that Maintenance, Quality, and Inventory are as important as Production itself and should work together with the same generic model, as described in Figure 16.2.

Before this model, the MESA model did not describe those other operations at this level of detail. The ISA-95 model provides widely accepted guidance for the manufacturing industry, so that consistent specification of requirements can be applied across all operations.

The ISA-95 model is functional and does not imply a specific given technology or architecture. As a reference model defined for integration purposes, the ISA-95 model allows various implementations of the MOM architecture without imposing a specific solution. It can be used to map how the enterprise information system is distributed in different applications or software packages, as described in Figure 16.3.

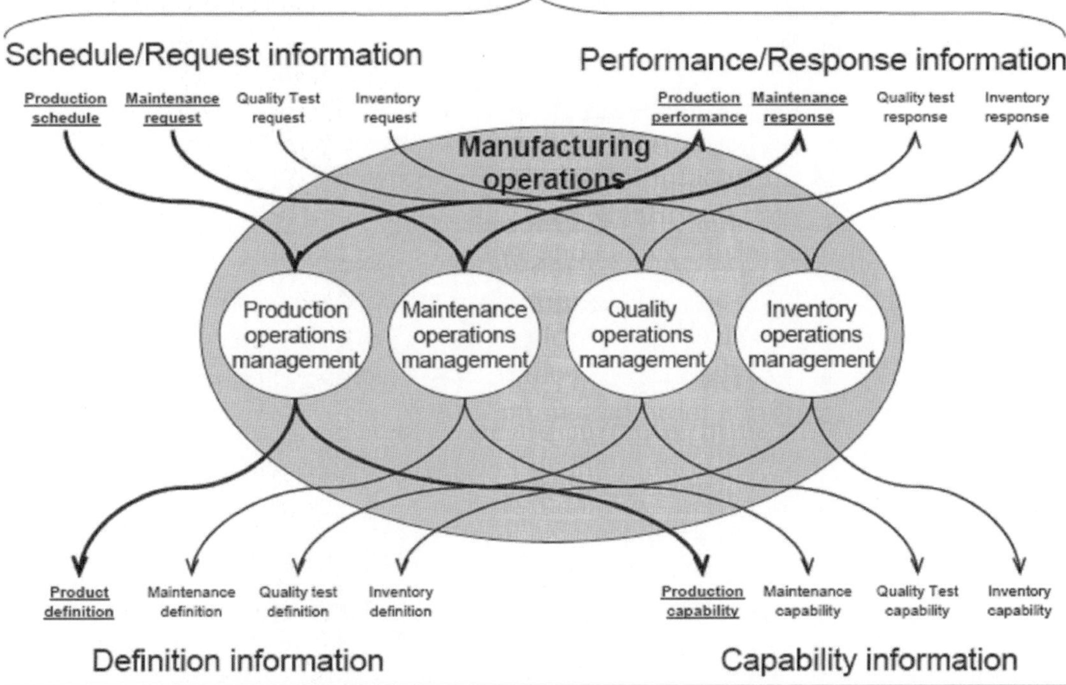

Figure 16.1. The ISA-95 part 3 MOM reference model and information flows with Level 4 PLM and ERP and Level 2 process control and automation.

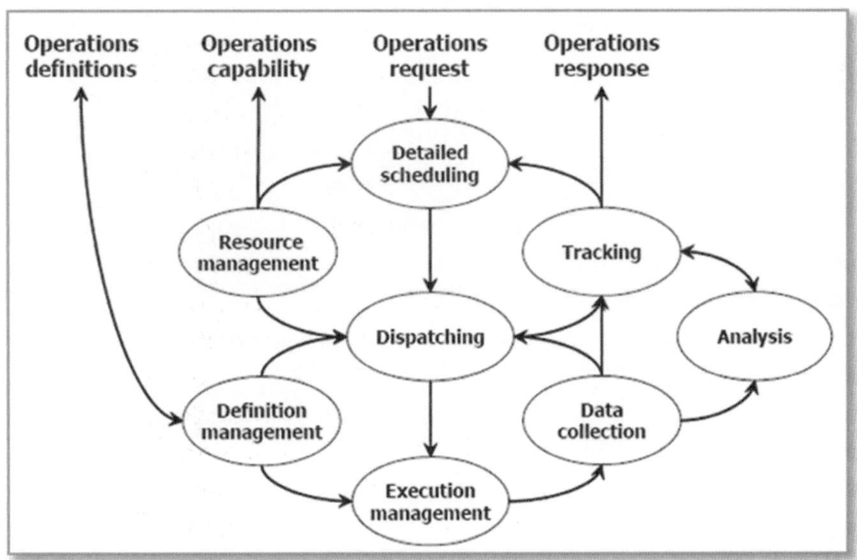

Figure 16.2. The generic activity ISA-95 model.

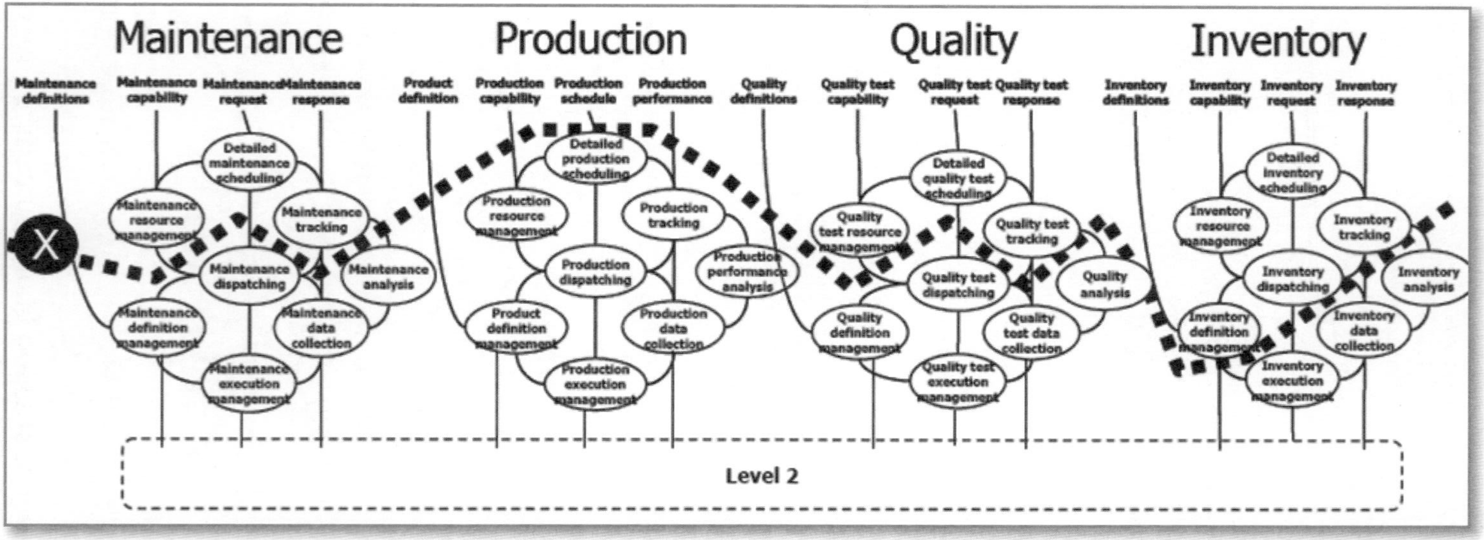

Figure 16.3. Use of ISA-95 model as information system function map.

Challenges of the ISA-95 Part 3 Reference Model

While this model is ideal as a reference or discussion model, from an architectural perspective it is not accurate to assume that any given manufacturing process could be supported with any given system.

The main problem comes from a lack of description of horizontal information flows; this model focuses primarily on describing the vertical flows with a department, business, or control level, often referred to as a "silo" of operations.

The generic representation of the activity model implicitly suggests that the underlying technology of software platforms could be common, or at least share some common components, in order to simplify and optimize the cost of implementation and Total Cost of Ownership (TCO).

Despite the vision of a balanced management of manufacturing operations, the reality of MOM architectures is still the result of a juxtaposition of legacy applications and stand-alone software systems, with little or no integration between their multiple databases (Fig. 16.4).

Implementing consistently integrated processes to support horizontal information flows is almost impossible in such architectures. This can cause severe limitation in implementing advanced processes such as Kanban-based production line replenishment, Just-In-Time (JIT) production, shipping, quality control in material warehouse receiving, and predictive maintenance processes using actual machine efficiency and production histories. For basic master data management, it is obvious that simple technical data such as product-related specifications need to be closely shared and synchronized between production, quality, and inventory.

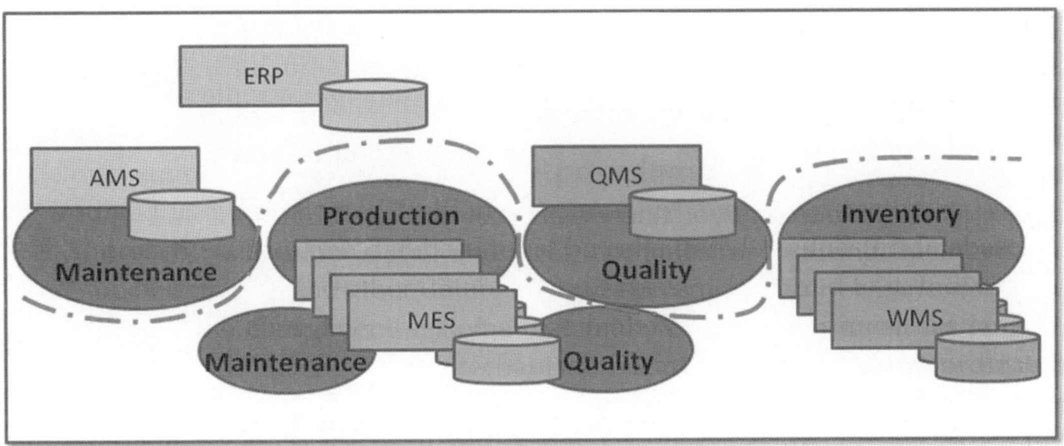

Figure 16.4. Today's MOM architecture made of complex combinations of local and central applications and databases.

The difference between a "best-in-class" production floor and a "laggard" performer comes from optimizing the manufacturing workflow with considerable information granularity. Add real-time information requirements to respond and adjust processes quickly, and you have a sophisticated requirement for a level of detailed information on individual critical resources within a highly dynamic environment, such as machines and tooling status. Work In Progress (WIP) levels, including the status of intermediate products and containers as well as their locations within the warehouse or the production area, are needed to effectively support intelligent decision making as unexpected events occur.

An inability to integrate disparate systems can lead to delays when customizing ERP-based manufacturing applications or when deploying and using MESs or WMSs across several sites, especially when these programs are really designed for a single location.

Multiple stand-alone software applications generate a complex IT infrastructure and a high cost of ownership. Very often, the number of these applications can be in the hundreds, severely limiting the ability to standardize best-practice processes as well as the ability to implement business process improvement for those plants that need it most (yet can afford it least)!

The Global Factor

In today's globally competitive environment, plants often require 24/7 availability to be able to compete—a requirement reinforced when production is globally distributed across multiple time zones. It is simply unacceptable to shut down a shop floor Operations Execution System (OES) for 2 hours for routine maintenance when operations run continuously. Somewhere around the world it will be in the middle of the day, causing extreme hardship for that region.

Add a real-time constraint of second or subsecond response times, and it is easy to understand the need for a local implementation of servers at the plant level with a decentralized architecture. Unfortunately, this decision raises challenging administration issues when operating globally distributed systems. Remote configuration and administration is needed, as most of the plants today have limited local administration resources or skills availability. Figure 16.5 illustrates some of the challenges that a global MOM solution must address, each of which is not specifically identified within the ISA-95 standards documentation.

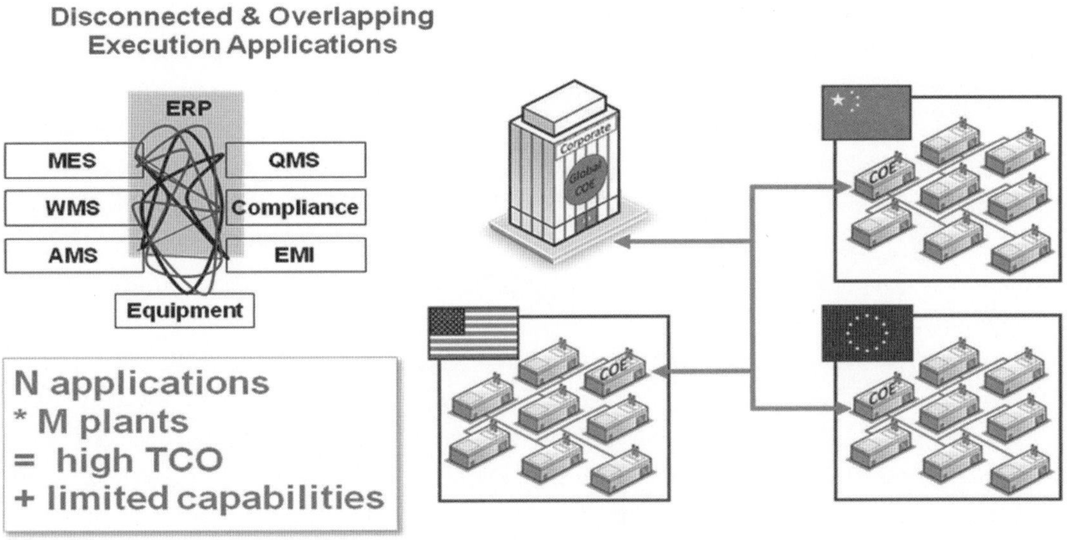

Figure 16.5. Complexity of multiplying and deploying MOM applications.

The Complexity Continues

The ISA-95 model also adds other activities required in manufacturing operations: security management, information management, configurations management, documents management, regulatory compliance management, and incidents and deviations management (Fig. 16.6).

When using multiple solutions for Production, Inventory, Quality, and Maintenance, it is necessary to generate multiple heterogeneous tools to manage each of these support activities, largely increasing the cost of implementation and ownership (Fig 16.7). Using applications developed by different companies results in the following:

- Heterogeneous architectures with a different look and feel for each user interface (that are harder to learn and teach)

- Multiple databases and data models, duplication, and management of master data

- Inconsistent process modeling (between routing or recipe, quality processes, and inventory management procedures such as receiving and put-away)

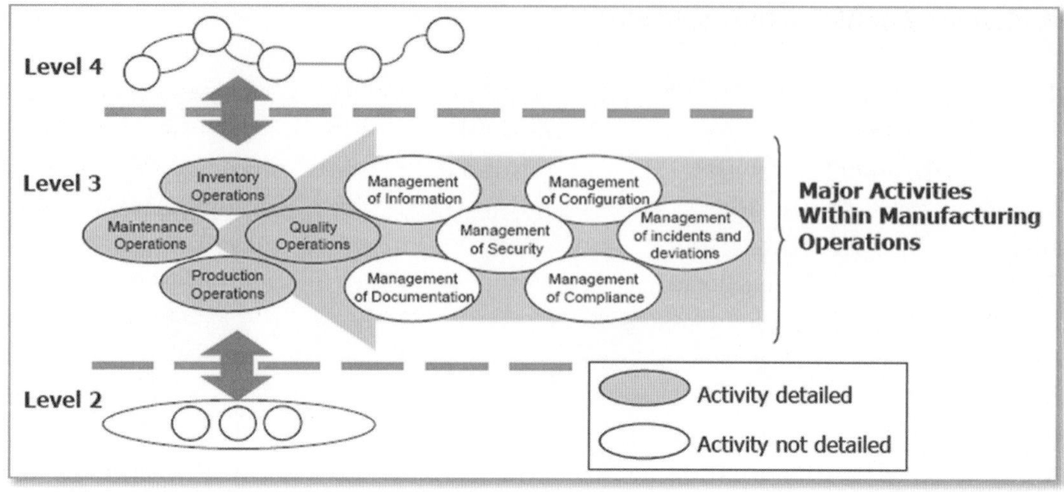

Figure 16.6. Additional common activities to support MOM.

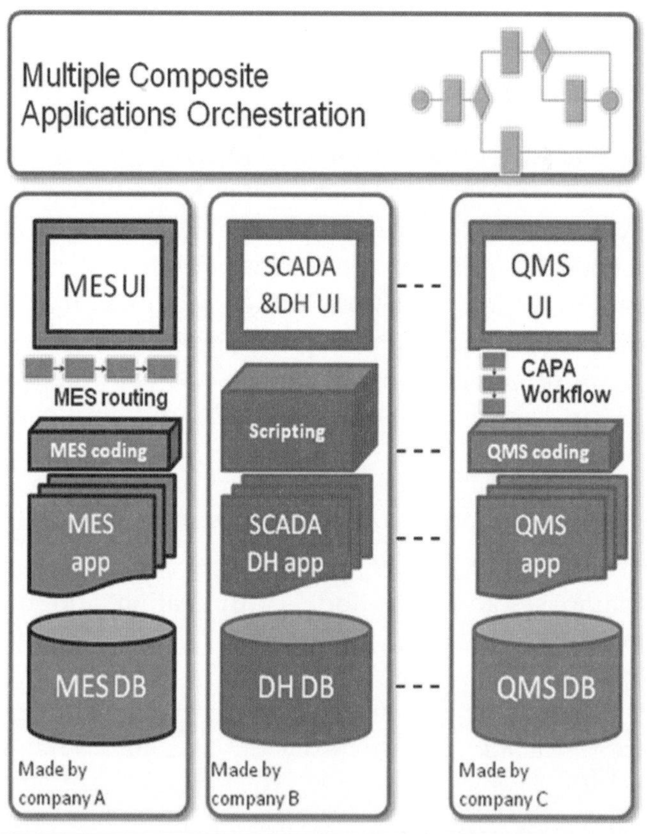

Figure 16.7. Heterogeneous software architectures for MOM applications.

SOA and BPM to the Rescue

The emergence of SOA and BPM has allowed manufacturing companies as well as software vendors to orchestrate and manage each of these various and heterogeneous applications by wrapping functions as Web services. But the complexity of implementation due to multiple data models and scripting technologies remains a limiting factor for large deployment of MOM applications, especially when distributed across multiple plants, geographic regions, and cultural divides.

Most MOM software packages (e.g., MES, QMS, WMS, Asset Management System [AMS], or ERP modules) were originally designed in the nineties based on client-server architecture. This means that functionalities are not configured with BPM technology but are instead traditionally coded. This precludes the ability to disassemble standard processes without custom coding. Despite the inclusion of a Web-based Graphic User Interface (GUI), very few of those applications are providing the capability to adapt a user interface without programming or scripting tools. And, last but not least, they do not cover the ISA-95 MOM reference model for operations execution with Production, Quality, Inventory, and Maintenance with the same unified architecture in various industries.

A Framework for Success

Both ARC Advisory Group and Gartner have introduced the idea of having three key enterprise platforms to support their business processes:

1. Enterprise, such as ERP, for financial and accounting, resource planning, and sales order management

2. Design, such as PLM, for product life-cycle management including product and process design and engineering

3. Operations, such as MOM, for manufacturing factory floor operations management

This logic follows a new way to look at and manage manufacturing, based on the simple idea that production is a process, not a set of isolated activities (Fig. 16.8). Everything in a plant ripples out and affects everything else. A parts shortage impacts the production line. A maintenance call affects labor availability and potentially product quality. A production delay changes shipping schedules.

Analysts and customers endorse the idea of a MOM—a consistent platform to support the definition and execution of any process spanning production, quality,

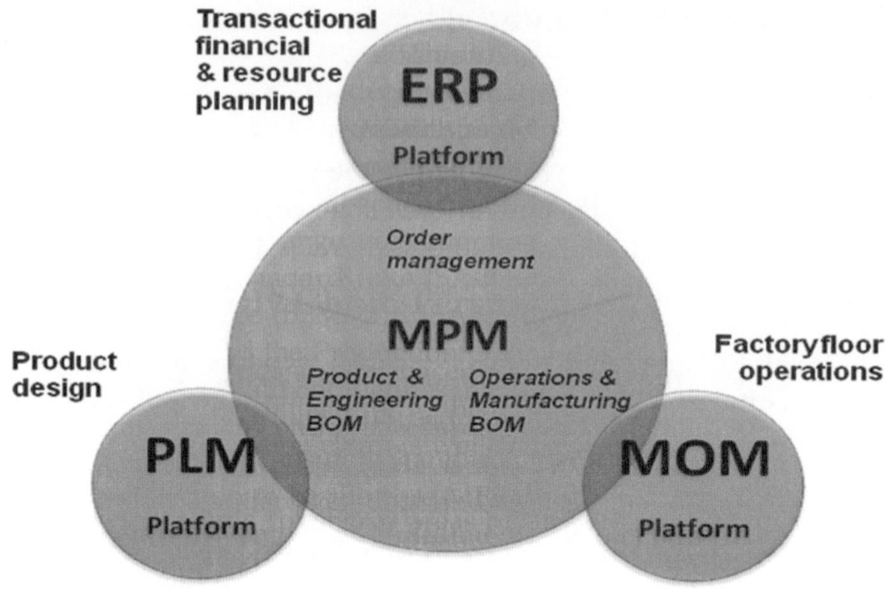

Figure 16.8. The three platforms: enterprise (ERP), design (PLM), and operations execution (MOM), integrated with Manufacturing Process Management (MPM).

maintenance, and inventory. Some would extend this concept into the supply chain as well, but that topic is beyond the scope of this chapter.

A possible solution is to use an OES architecture designed to consistently support the ISA-95 model for all manufacturing operations through the following (see Fig. 16.9):

- Single data model

- Single manufacturing process model

- Single Web-based GUI

To be as successful at the shop floor level as ERP has been at the back office level, MOM needs to provide the following:

- One shared database for a "single version of the 'plant' truth"

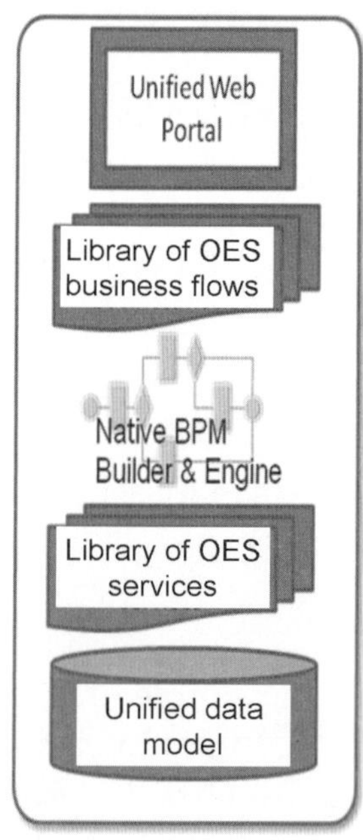

Figure 16.9.
An OES architecture.

- Cross-functional and cross-departmental corporate manufacturing processes (Production, Quality, Inventory, and Maintenance)

- Enterprise-wide standardization of processes for business units or geographies within select manufacturing departments

Three examples are highlighted in the following sections that show how an OES with a unified architecture has been used in Aerospace and Defense (A&D), automotive, and consumer goods.

SOA and BPM Applied in A&D Manufacturing Operations

A worldwide leader in the defense industry with production in three sites in different countries is producing complex missile systems with a high level of customization. Products and assembly processes are defined using CAD Pro-engineer and Teamcenter PLM systems. Planning is done using SAP R/3 ERP & SCM. Apriso FlexNet was selected as the MOM for process and work instruction authoring and production execution. The challenge consisted of the following factors:

- The need for very detailed step-by-step work and quality control instructions requiring significant preparation by process engineers

- Very high quality standards and process traceability

- The need to replace an obsolete authoring application not supported and not connected to SAP

- The need for connectivity to specific equipment such as screwdrivers, scales, and bar code readers

The workflow is the following:

- "As-designed" product and process data (e.g., components, Bill Of Materials [BOM], routing) are produced with the PLM

- The routing becomes the process version 0 after import in the MOM

- Authors add work instructions, link pictures and documents, perform Quality Assurance (QA) and error proofing, and allow traceability with the process authoring tools in the MOM

- "As-planned" process orders, with the specification of some serial components to be used from the inventory, are sent by the ERP

- The MOM generates the tasks and drives operators step by step for the assembly process execution and generates the "as-built" traceability by collecting data from machines

- The "as-executed" production progress and quality results are sent back from the MOM to the ERP

The manufacturing processes are consistently defined and transferred in the PLM and then enriched, executed, and traced in the MOM with the ability to cover both manufacturing and quality activities using the same process model, sharing the same database and the same user interface.

SOA and BPM Applied in Construction Equipment Manufacturing

Volvo CE is a leader in construction equipment and operates on a worldwide basis through a network of twelve plants on four continents. Volvo initiated a global effort to increase operational efficiency and competitiveness, starting with the replacement of their existing MES implemented in Korea and with the goal to better extend the value of the SAP deployment. The selected approach was then to extend the initial functional scope to create a common core system that would apply to all manufacturing sites. The objective for the plant was to build a MOM to support their "Manufacturing Excellence" targets. For Volvo IT, the strategy was to have a software platform that was flexible, deployable, well integrated within the SAP Factory Master Program, and cost effective to maintain. The Volvo CE group expected to obtain an information system that would allow the leveraging and sharing of best industrial practices among the plants.

The MOM functional scope covers production control with the management of all sequencing of orders sent by SAP, the tracking of all orders, the monitoring of cycle time and performance, real-time visibility on alerts coming from the assembly lines, and management of necessary documents needed for the production of bar code labels and body build orders.

The same MOM also covers labor management, with collection of attendance data for performance indicator calculation; logistics execution, managing all production line replenishment with line feeding lists for the logistics operators; and quality, with the collection of all data and the tracking of all in-line inspections and audits.

This is another example of a consistent application supporting ISA-95 Production, Quality, and Inventory using one single BPM process modeling technology, a unified data model, the same Web-based graphical user interface, and SOA integration with the ERP.

SOA and BPM Applied in Consumer Packaged Goods Manufacturing Operations

L'Oréal is a well-known worldwide leader in cosmetics, producing a very large family of products using batch manufacturing processes. L'Oréal selected SAP and Apriso to build their new manufacturing information system called ISIS, using a central SAP system for supply chain management and ERP and the distributed architecture of Apriso for its MES, Quality Execution System (QES), and Logistics Execution System (LES).

The challenge was to build and deploy this global application in about twenty-two plants in a short timeframe, replacing a homemade legacy system that was covering ERP, MES, QES, and LES. The manufacturing covers weighing and dispensing, batch processing, packaging, and palletizing. The material flow is organized in a very Lean way, from the raw material inventory up to the shipping of pallets of finished goods with as little WIP as possible. Technical product data come from SAP (e.g., products and materials, formulas, characteristics, specifications, inspection plans), while recipes are detailed in the MES system.

As in this Consumer Packaged Goods (CPG) example, more and more customers seek to consolidate and rationalize their IT infrastructure at the plant and warehouse level. Many hoped that SAP would do it all but are increasingly concluding that ERP cannot do everything. They want to minimize the number of other applications they install and implement. All activities for Production, Inventory, and Quality are supported by the same execution solution. In the case of L'Oréal, this was needed to make sure that integrated collaborative processes were sharing information in real time, as follows:

- Weighing process needed to display to weighing operator status and bin location of material in inventory during the weighing process
- Production (i.e., mixing and packaging) needed to have access to material inventory for production line replenishment
- Quality control needed to share information with material inventory management for receiving inspection
- Quality control also needed to share information in real time with production to dynamically calculate sampling quantities based on severity (per ISO 2859)

A full Microsoft platform including an SQL server was finally selected to minimize TCO, knowing that limited IT resources were available on-site.

Toward Global Manufacturing Operations Management

There is a critical commonality between the previous examples: the need for consistent manufacturing process management covering the MOM functional scope, as defined by the ISA-95 model in terms of Production, Quality, Maintenance, and Inventory. For a multiplant company, this allows an MOM skills center to model, configure, and deploy processes consistently. Leading manufacturing enterprises are focused on capturing and deploying best practices throughout their organizations. This approach helps maintain and improve quality standards, consistent operations, and measurable performance metrics.

One such approach that effectively addresses systems integration challenges is to complement an ERP deployment with an OES and deploy them synchronously (thus embracing a "core" systems implementation approach). The focus then becomes to identify the approximately 80% of business processes common to each plant, which can then be supported using a standard configuration of the ERP and OES software at each location. The OES can still provide the flexibility to configure the remaining 20% of the manufacturing processes requiring local customization. This type of OES supports managing processes *globally* and executing *locally*.

Global process management means that as processes are defined, configured, and tested in various locations, they should then be stored in a central repository. It must be possible to transfer these processes to each plant location to perform the required local execution.

Surprisingly, this transfer capability, based on strict versioning and configuration management, is not offered as a standard feature in most MES products. It would appear that few vendors have seriously considered the challenges of a global rollout. Another benefit to such an approach is the ability to govern process changes, while documenting that the appropriate approvals are in place to authorize any change.

These capabilities need to include not only the entities as processes and operations, but also all possible linked entities that are part of an execution task. Such a system must be capable of transporting various formats of configuration objects, such as database entities, Extensible Markup Language (XML) files, program files, and so on.

This is fundamentally different than simply standardizing on an MES platform and developing one application for each plant, leading to a 6- to 12-month project implementation time per plant. With a core implementation, the deployment of the core profile for the MOM application can be installed in as little as 30 to 90 days, depending on the scope and complexity.

Further Reading

Apriso. 2007. The need for SOA to achieve flexibility and agility within plant operations. White paper. http://www.apriso.com/library/white_papers.php.

Apriso. 2008. Core deployment program for enterprise operations execution. http://www.apriso.com/services/global_core_deployment.php.

ARC Advisory Group. 2006. Collaborative manufacturing strategies: The critical role of plant IT. Presented at MESA conference, October 2006.

Hughes, Andrew, and Marc Halpern. 2007. Manufacturing process management enhances workflow across PLM, ERP, and MOM. http://www.gartner.com/DisplayDocument?id=509664.

IndustryWeek. 2008. Deploy best practice processes across the global enterprise: Featuring a case study by L'Oréal. http://www.apriso.com/library/webcasts_archive.php.

Manufacturing Business Technology. 2007. To reach desired "IT end-state," Volvo CE focuses on multi-plant performance. http://www.mbtmag.com/article/CA6487608.html?nid=3679 (accessed October 2007; site discontinued).

Best Practices for MES User Requirement Specifications

Presented at the WBF European Conference, November 10–12, 2008, by

Bianca Scholten
Management Consultant
bianca.scholten@ordina.nl
TASK 24
Burgemeester Burgerslaan 44
5245 NH Rosmalen, The Netherlands

Abstract

System integrators start their Manufacturing Execution System (MES) implementations by writing Functional Specifications (FS). The input for these FS is the User Requirements Specification (URS) document. This URS is written by the end user, often with the help of external consultants. When comparing URS documents from different sources, there appears to be many differences, especially in terms of scope, contents, and level of detail.

The technical consultants who write the FS need certain input at a specific level of detail. Also, in order to be able to make a quotation for the MES project, specific information is needed. So it is important for the authors of the URS to understand what kind of information they have to deliver to the supplier (i.e., technical consultants and sales people). However, different parties seem to have different opinions about the types of information that URSs should contain.

A thorough comparison of some URS documents from different sources was done in order to document best practices and standardize internal processes. This chapter reveals best practices and pitfalls to avoid, in order to improve the "handshake" between the URS phase and the FS phase of MES projects. The final goal is to improve communication between end-user companies and MES suppliers and system integrators.

Introduction

Being a "Client" Is Not Easy

Last year I bought a house. It was perfect, except for the bathroom. That summer I went to the showroom of a well-known bathroom supplier. "Would you like some coffee?" the lady behind the desk asked me. I responded, "Yes, please."

I was sure that buying a new bathroom would take a lot of time, so I could use a cup of coffee. But to my surprise, I ended up all alone with my coffee at a table. "This is not working," I thought. I waited until I saw the lady again and then raised my hand. "I would like to talk to someone about bathrooms."

A few moments later a young man came toward me. I presented myself and said, "I want to buy a new bathroom." He stared at me for a while, until he asked, "And you would like to make an appointment to discuss that?" "No," I responded. "I would like to discuss it right now!" "Oh, well . . . That's possible . . . Please take a seat while I get my materials."

He came back with some paper, a box of colored pencils, and a ruler. I told him the size of my bathroom, the locations of the window and the door, and that I wanted a bath, a shower, a toilet, and a vanity. He made a few drawings of possible compositions. After that, we walked through the showroom, and I pointed out the bath, shower, and all the other things that I wanted to buy. "When will the quotation be ready?" I asked. He replied, "You will receive it within a week, by regular mail."

One week later I indeed received an envelope. The salesman had enclosed the drawing and a price: 7500 euros. Nothing else was in the envelope. I decided to call him.

> "Thank you for the quotation that I just received, but you forgot to enclose a specification list. And moreover it doesn't say if the price includes the installation."

> "It always excludes installation. We never install the bathrooms ourselves."

"OK. But could you please send me the list of specifications of the things that I chose?"

"We never do that. We purchase the parts ourselves and we can gain a margin on that."

"But then how can I be sure that you are going to deliver the things that I chose? And what if I find it too expensive? How can I know what the most expensive parts are in case I want to choose cheaper ones?"

We did not come to a conclusion. The guy kept on saying that this was their usual way of working. I put down the telephone. I am not going to buy a bathroom over there!

In our company, my bathroom story has become a frequently used metaphor for bad communication between a vendor and their client. As a management consultant, I help industrial companies describe their requirements for new MES systems. The deliverables are a list of requirements, plus a textual document providing contextual information about these requirements.

How can a client describe their requirements in a clear and unambiguous manner? And how should a vendor specify what exactly they are going to deliver? Together with my colleagues from our MES and ISA-95 competence center, we decided to establish an internal working group and look for best practices.

URS According to GAMP

First, the working group needed to agree on the naming of each phase in the purchasing and implementation of an MES—especially the phases in which a client describes their requirements and the phase in which the supplier explains what they are going to deliver. (This was the scope of the working group.) MESs are a rather new concept, and there is no standard project method available for implementing these systems. Thus there is no standard definition of the MES project phases either. What names are we going to give to these phases? Some people in our team suggested picking names from the ERP world. Others came up with names from the military. But, in the end, our consultants appeared to be very much in favor of using terminology based on Good Automated Manufacturing Practice's (GAMP's) V-model, which in itself uses terminology from the IEEE 12207 standard.

What is GAMP? For many years, GAMP has been an accepted methodology to realize validated control systems. Industrial companies that have to comply with FDA regulations must deal with inspections. During these past inspections, the FDA saw that certain companies had a good way of working, and they wanted the other companies to work at least as well as these companies. Thus the best way of working of specific enterprises became the standard way. The FDA's requirements became increasingly stricter.

GAMP is an initiative of the International Society for Pharmaceutical Engineering (ISPE). The ISPE is a community for the pharmaceutical industry. They developed GAMP in order to have one common, standard interpretation of the highly abstract GxP regulations. Note that "GMP" is obligatory, whereas GAMP is a methodology that may or may not be used. The V-model from GAMP has become very famous (Fig. 17.1).

GAMP also has disadvantages. It is especially suited for enterprises that have to comply with FDA regulations. So it is a rather strict method with many extensive documentation and qualification steps. For companies that do not need to comply with FDA regulations, all this specifying and testing seems a bit overdone. It takes a lot of time. A new phase may only be started when the total previous phase has been completed and Quality Assurance (QA) has given approval. Both forward and backward traceability are required. When, during the building phase, it is discovered that the requirements in the URS are not complete or are wrong, the whole trajectory (i.e., URS, FS, Design Specifications [DS]) has to be redone.

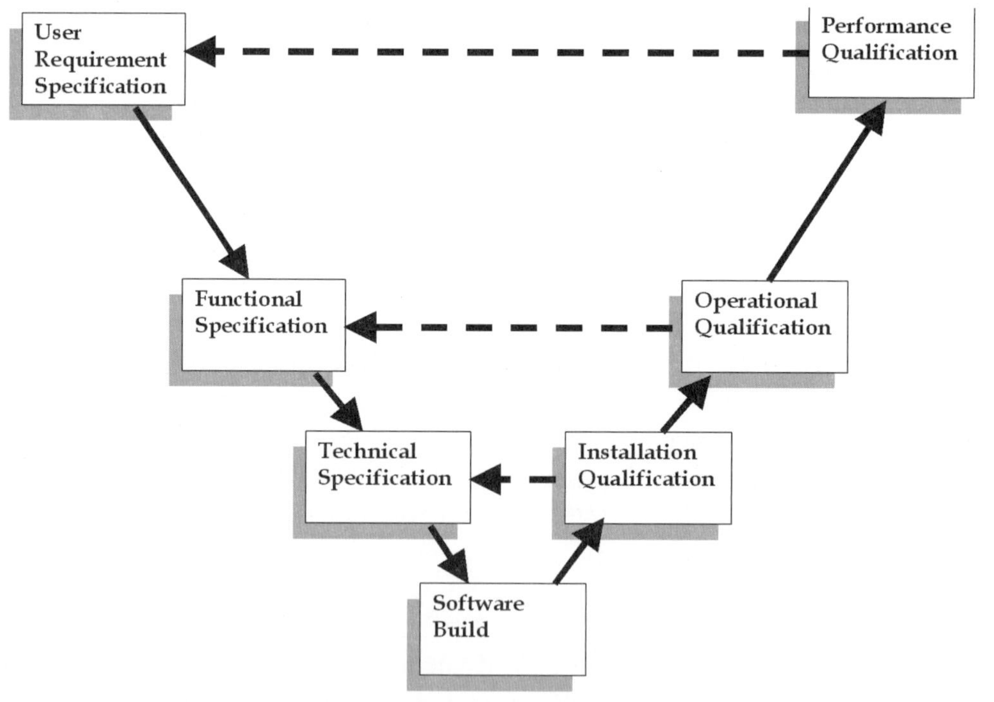

Figure 17.1. The V-model of GAMP.

Despite these strict rules, we decided to choose the GAMP V-model and its terminology to define our project phases. Unlike the other suggested terms (from the ERP world and the military), GAMP was especially developed for industrial processes. And in case an industrial company does not need to comply with FDA regulations, then they can always use the "GAMP light" version (meaning they only use GAMP phase names and terminology without following GAMP's strict rules for validation and qualification).

GAMP's famous V-model (Fig. 17.1) makes it clear that automation projects contain different phases. During the first phase, the client specifies the user requirements (meaning that the employees of the manufacturing company decide what they want, often with the help of an external consultant). Note that a URS is not only a document that explains to the supplier what is expected from them; it also reports the important decisions that users took and reflects the internal agreements of the end user's company. During the URS phase, the future users become conscious of their wishes and requirements for the future situation.

After the client has written the URS, the supplier can start to write the FS. In the FS document, the supplier tells what the final system will be able to do (i.e., what they are going to deliver). The FS is a response to the URS. The URS answers the question "What?" and the FS answers the question "How?" The URS is described independently from the future solution. The FS describes the solution. The FS is related to a requirements traceability matrix document to assess if all the user requirements are met.

If the supplier is not able to provide a solution for a specific user requirement, then he has to explicitly state this. The FS has to be written in such a way that the client is able to understand it. (Note that in most cases the client is neither a technician nor a programmer.)

On the one hand, the FS is a means to communicate with the client. But on the other hand, it is the input for the next phase, in which the supplier writes the DS. The DS contains a detailed, technical description of the solution. This does not need to be in a language that the client can understand, because it is not meant to be read by the client. The DS is read by people who are going to build and implement the solution (i.e., the supplier's engineers and programmers). After building the solution, many qualification phases follow, but these were all out of scope for our working group. (This chapter focuses specifically on the contents of a URS). From the previous insights, it can be concluded that the URS has two important goals:

1. Report the internal decisions of the customer about the functionality and other aspects of the new system

2. Inform the supplier about what is expected from the solution that he is going to deliver

What the V-model of GAMP does not make clear is that, in between specifying the user requirements and the functional requirements, often a solution selection and a request for quotation take place. This is an important fact that highly impacts the required contents of a URS.

Comparison of Real-life URS Examples

Most authors of URS documents have a clear picture of the goal of the URS. Nevertheless, when comparing the URS documents written by different authors (even authors from the same company), there appears to be huge differences in scope, content, and level of detail. Our consultants came to the conclusion that it was necessary to develop a URS template based on best practices. This could lead to the following advantages:

- We would be able to explain the deliverables of the URS phase to our customers in more detail. (Many of our consultants guide clients during the URS phase, so during this phase the client and the consultant need a clear picture of the goals and deliverables.)

- We would have a guideline to deliver high-quality URS documents.

- We would have a tool to train our junior consultants.

- We would be more certain that we offer the supplier (in some cases, our own colleagues) the right amount of information at the right level of detail to serve as an input for making quotations and writing FS documents.

- Our sales people and technical consultants would have a means to assess URS documents received from clients. If these do not provide all the required information to be able to make a quotation and write an FS, then they would be able to explain in detail what information is lacking. Thus the template would become a means to decide on a "go" or "no go" for the next phases.

In order to develop a URS template for MES, we compared four examples. Two of the examples were written by clients and sent to our company to make a quotation and write an FS. Two other examples were written by our own internal consultants for two different clients. After the URS phase, these clients asked our company for a quotation, and after that we did the implementation, starting with writing an FS. So in all cases, our sales people and technical consultants were able to tell us if the URS provided enough information and if it provided the required level of detail to be able to make a quotation and write an FS.

Note that the Joined Equipment Transition Team (JETT) of the ISPE has also developed a URS template. We took a close look at this template and decided that it was not fit to serve as a basis for a URS for MES. (It completely focuses on production equipment, including the related control systems.) So we only used this template to perform a final check and see if we had not forgotten important subjects. We compared the table of contents of the four examples plus the level of detail. We encountered many differences, including the following:

- *Scope of the document.* One of the documents enclosed a lot of information for a request for quotation (e.g., request for information about the system integrator's employee skills), whereas the other three documents did not.

- *Amount of pages.* This varied between 50 and 110 (including annexes).

- *List of requirements.* Our own documents included an annex with a list of all requirements, whereas the others only had textual descriptions of the requirements without listing and numbering them (tagging).

- *Contents.* Some of the documents included information about hardware requirements, whereas others did not.

There also was one important similarity: all the examples used ISA-95 terminology, structures, and models.

Conclusion: URS Template for MES

The comparison led to an MES URS template. The appendix of this chapter contains an overview of the table of contents of the template, plus the advised level of detail per chapter and paragraph. There are some points of concern and some pitfalls to avoid when applying this template:

- The ordering of the contents can be done in a different way, as long as no essential information is left out.

- The suggested contents are based on the idea of a broad project scope. Subjects that are out of scope in a specific project should be deleted in the actual URS. For example, if the client only wants to purchase reporting functionality, then subjects like production resource management and product definition management may be left out.

- In the basic examples, sometimes the wording "Functional Specifications" was used as the title of a paragraph or chapter within the URS. This is confusing because GAMP uses this terminology to point at a document that should be written in the next phase. In the template that we propose, the use of this wording is avoided.

- The URS should be a combination of a numbered list of requirements with a document that provides extensive contextual information about the requirements. Always add an appendix containing the requirements list! The list of numbered requirements is the basis for the supplier to make a cross-reference table that explains which functional requirements meet which user requirements. Note that there are books available about best practices in phrasing requirement sentences (see the further reading list at the end of the chapter).

- Graphs, models, flowcharts, and so on make it easier to understand the textual descriptions.

- If currently an MES system is in use that is going to be replaced by a new system, then add a chapter to describe the functionality of the old system, including its strong points that the new system will need to be able to do and any issues in the old system that will need to be avoided in the new system. Describe the functionality of the old system using ISA-95 terminology and text structures. Merge the resulting requirements into the final requirements list in such a way that the reader does not have to be familiar with the functionality of the old system.

- If a Maintenance Information System (MIS), Laboratory Information Management System (LIMS), or Warehouse Management System (WMS) is required, then paragraphs can be added in the same way as Production Operations Management for the following:

  - *Maintenance Operations Management.* Add paragraphs for maintenance definition management, maintenance resource management, detailed maintenance scheduling, maintenance dispatching, maintenance execution management, maintenance data collection, maintenance tracking, and maintenance performance analyses.

- ❏ *Quality Test Operations Management.* Do the same for quality.

    - ❏ *Inventory Operations Management.* Do the same for inventory.

- Explanations of standard terminology (e.g., ISA-88 terminology and ISA-95 terminology) should be avoided. Just refer to the standards, where people can read the meaning of this terminology. In an annex you can add a short description of the purpose, contents, and most important models and terminology of the standards.

- Describing user requirements in terms of solutions should be avoided. It is up to the vendor to think of a solution to the client's problems.

- The ISA-95 models and terminology provide an excellent basis for the URS template, but the scope of ISA-95 is broader than MES. ISA-95, on the other hand, does not provide models for all the information that is required in a URS for MES.

- Separate documents should be written for subjects that do not belong in a URS (e.g., requirements for the system integrator and Request For Information [RFI] and Request For Quotation [RFQ] information). Do not include these in the URS.

- The proposed template provides paragraphs that describe the "As-Is" situation and paragraphs that describe the "To-Be" situation. Note that it is also possible to have two separate chapters, one for the As-Is situation and one for the To-Be situation. It is a matter of taste if you combine or separate this information. The important thing is that the information is available somewhere in the document.

A Happy Ending

By using this template for a URS for MES, you can avoid scenarios like my bathroom story. Do you want to know how that ended? I went to another supplier who not only sold bathrooms but also installed them. The saleswoman did not use colored pencils nor did she make a fancy drawing. She informed me that if I chose that specific vanity, I'd need an installation set that cost 15 euros. She also specified that the particular tub I chose required a set of supporting legs that cost 30 euros. She took two hours to help me specify the complete list. Then we talked about possible savings. For the most expensive parts, I decided to choose an alternative solution. The saleswoman advised me to choose the accessories only after the installation. In her experience, clients often change their opinion after the bathroom is ready.

So we decided to make a rough estimation for the accessories and leave them out of the fixed price.

A few months later, I got the key to my new house, and they came to install the bathroom. Two men removed all the existing devices and tiles. Then came the electrician. He brought the basic sketch that the saleswoman had made. "You told our colleague that you want electricity near your washing stand. Where exactly should that be?" I pointed out the location. Based on his experience, he advised me to do it differently. I agreed. The next day, two men came for the tiles. They also called me when they were not sure about details on the sketch. After two weeks, the work resulted in a beautiful bathroom that was professionally installed, delivered on time, and within budget.

To me this is the perfect example of good communication between a supplier and a client. Make a requirements list that provides information that is detailed enough. Be open and clear about costs and efforts that are not obvious. And don't expect the requirements list to answer all the questions. Keep talking with the client during all the phases! Keep in close contact with the client and use the URS document as the basis for the communication.

Further Reading

Scholten, Bianca. 2007. *The road to integration: A guide to applying the ISA-95 standard in manufacturing*. Research Triangle Park, NC: ISA.

Dijkgraaf, Wim, and Mike van Spall. 2007. *Start at the end with SMART requirements*. Netherlands: Synergio B.V.

Appendix

Chapter	Paragraph	Subparagraph	Explanation plus advice for the level of detail
	Current application architecture and infrastructure		List applications and describe the architecture of hardware, software, networks, and bus systems in place. (ISA-95 models can be used as a checklist.)
	Users		Provide a short description of the roles of the future users of the system. Also, provide a basic description of how they'll use it.
Business drivers			Describe drivers and business goals (from input from interview with higher management). See Annex B of ISA-95 part 1 version 1. Describe the impact of the business drivers on the MES requirements. Make clear what is the (high-level) business case for the MES system. In other words, define the expected advantages in terms of cost savings, improved customer service, higher flexibility, and so on. (Please note that at this stage it is usually not possible to provide detailed, exact figures about cost or time savings.)
Production Operation Management	Product Definition Management	Product Definitions	Give representative examples of product definitions (e.g., recipes, assembly instructions). Describe complexities like routings, dependencies, and master data synchronization. Give the number for the approximate total amount of recipes. Describe which process segments (lines) are used to execute different groups of recipes. Describe the product segments from a Level 4 (ERP) point of view (e.g., intermediates and finished products for which the ERP provides a Bill of Materials [BOM]).

Chapter	Paragraph	Subparagraph	Explanation plus advice for the level of detail
		As Is	Describe current way of working (development of new product definitions, downloading of product specific settings to Level 2, current systems, points of concern, etc.). Describe flexibility (e.g., for new product introductions). Describe version management or product instructions or recipes. Describe automation level.
		To Be	Describe future situation in the same level of detail.
	Production Resource Management	Production Resources	Give typical examples of production resources (e.g., equipment, material, and personnel and their properties) as relevant for the new MES. Provide an estimated total number of resources.
		As Is	Describe current way of working (i.e., the way that information is maintained and provided about the availability of the resources and their capacities and skills).
		To Be	Describe future situation in the same level of detail.
	Detailed Production Scheduling	Production Schedule	Give typical example of the information in a production schedule that is received from Level 4, plus a detailed production schedule that is made within Level 3.
		As Is	Describe current way of working (i.e., the way that a schedule is received from Level 4 and the way that a detailed schedule is developed on Level 3). How often is the schedule received? How is feasibility of the plan communicated? What is done in case of changes to the schedule?
		To Be	Describe future situation in the same level of detail.

Chapter	Paragraph	Subparagraph	Explanation plus advice for the level of detail
	Production Dispatching	Production Dispatch List	Give a typical example of information in a dispatch list, if this exists in the company.
		As Is	Describe current way of working. Describe the way that the orders are dispatched to production execution management. Describe if this is automated, done on paper, or done by phone call, and so on. Describe what kind of information is provided to whom.
		To Be	Describe future situation in the same level of detail.
	Production Execution Management	As Is	Describe current way of working for operators and supervisors during production, from receiving the order and product definition information (e.g., recipe) to finishing the order, including handling of errors and events. Describe this per work center within scope (e.g., per production line or process cell). Provide a detailed definition of the steps in the interaction of the operator with the system, per work center. Describe the exact information that the system has to provide to the operator during production. Describe the exact information that the operator has to enter into the system and how this is done.
		To Be	Describe future situation in the same level of detail.
	Production Data Collection	As Is	Describe current way of working. Describe manual and automatic data collection. Describe per work center (e.g., per production line or process cell) the data that are collected (e.g., charged amounts, temperatures, machine break downs, operator comments, etc.).

Chapter	Paragraph	Subparagraph	Explanation plus advice for the level of detail
		To Be	Describe future situation in the same level of detail.
	Production Tracking	Reports	List reports within scope and describe their contents.
		As Is	Describe current way of working. Describe manual and automatic generation of reports. Describe the situations in which each report is used (e.g., at a change of shifts). Describe who uses the reports and for what purpose.
		To Be	Describe future situation in the same level of detail.
	Production Performance Analysis	As Is	Describe current way of working, analysis tools in place, issues, points of concern, and so on. Also describe the feedback of production information to Level 4 (e.g., information for costing and information about actual materials consumed).
		To Be	Describe future situation in the same level of detail.
Interfaces	Interfaces with Level 4	(List the ISA-95 information flows)	See ISA-95.01, which describes thirty-one information flows. For each information flow, determine if it is within scope and decide if the interface will be automated. For each interface to be automated, give detailed information about source and target application, data within the interface, trigger (i.e., when is the information sent), data types, interface technology, and other requirements.
	Interface with Maintenance Operations Management		Describe how production and maintenance exchange information, and in case an automatic interface is required, describe details of this interface (e.g., data in the interface, trigger, source and target system, etc.).

Chapter	Paragraph	Subparagraph	Explanation plus advice for the level of detail
	Interface with Quality Test Operations Management		Do the same for the quality test (e.g., interfaces with the LIMS).
	Interface with Inventory Operation Management		Do the same for inventory (e.g. interfaces with the WMS).
	Interfaces with Level 2		Describe the required interfaces with Level 2, including source and target applications, data within the interfaces, triggers, required interface technology, and so on.
Other activities	Management of Information		Describe requirements for information management (e.g., information storage, back ups, recovery, redundancy, archiving). Refer to relevant internal or international standards that have to be followed. Describe the estimated amount of data (e.g., the amount of recipes or amount of material definitions within scope). Describe data types.
	Management of Documentation		Describe requirements for management of documents (e.g., the requirements for MES functionality to deal with Standard Operating Procedures [SOPs], recipes, batch records) and the procedures in place to manage corporate documents, including disaster recovery.
	Management of Security		Describe requirements for security management, like authorization possibilities per role, security concerns in case of remote access, and so on. Describe requirements concerning passwords. Describe required access levels. Provide a general description of the access rights for each level.

Chapter	Paragraph	Subparagraph	Explanation plus advice for the level of detail
	Management of Regulatory Compliance		Describe requirements for regulatory compliance that impact the MES (e.g., requirements for electronic records and signatures, SOX requirements, FDA requirements, etc.).
	Management of Configuration		Describe requirements for configuration management and change control procedures that are within scope of the MES, including requirements for audit trails and revision management, version management, master data management, and so on.
	Management of Incidents and Deviations		Describe requirements for incidents and deviations management (i.e., requirements for recording of incidents), in case this is a requirement for the MES.
Miscellaneous	Users, Workplaces		Describe user roles, amount of users, amount of workplaces, and locations of the workplaces. Provide details of the physical environment in which the system will be operated.
	System Administration Requirements		Describe requirements from the point of view of system administration.
	System Documentation Requirements		Describe requirements for system documentation.
	General System Requirements		Describe system performance requirements, database requirements, system uptime and response time requirements, availability, hardware and software requirements, clients and servers, Graphical User Interface (GUI), scalability, flexibility requirements, preferred standards (e.g., ISA-88, ISA-95), and so on.
	Language		Describe requirements concerning language.

Chapter	Paragraph	Subparagraph	Explanation plus advice for the level of detail
Next steps	Roadmap		Describe priorities of groups of requirements (e.g., "in the first phase only tracking functionality will be implemented"). Describe the roadmap for the implementation of all requirements. Also pay attention to the required training and coaching of users to move from the current situation to the future situation.
	Project Method		Describe the project methods that are used, project phase terminology, project roles, and so on. Give short explanations of the meaning of terminology that is typical for the chosen project method.
	Project Schedule		Describe project phases (e.g., RFI, vendor selection, RFQ, FS, DS, build, etc.), including detailed information about deadlines for phases in the near future plus rough estimates of deadlines of the later phases. Describe early phases in more detail; describe later phases in less detail. Describe milestones and timelines and responsibility of vendor versus responsibility of end user.
Approval			Insert your company's standard approval page at the appropriate position in the document.
Appendix A: Requirements List			Include a numbered list of requirements. For each requirement there has to be a reference to the paragraph(s) in which contextual information is provided about that requirement. Describe requirements in a "SMART" (Specific, Measurable, Attainable, Realistic, Timely) way.
Appendix B: Examples			Provide examples of recipes, flow charts, reports, schedules, and so on.
Appendix C: ISA-95			Provide a short explanation of ISA-95.

Chapter	Paragraph	Subparagraph	Explanation plus advice for the level of detail
Appendix D: Definitions and Abbreviations			Provide definitions and explain abbreviations.
Appendix E: Reference Documents			List reference documents.
Appendix F: Legal Aspects			Provide legal documentation (e.g., a nondisclosure agreement form).
Appendix #			Add any required appendices.

CHAPTER 18

ISA-95 Applied as an Analysis Tool

Presented at the WBF
European Conference
November 13–15, 2006, by

Bianca Scholten
Partner
bianca.scholten@ordina.nl
Ordina ISA-95 and MES competence centre
Science Park Eindhoven, Son,
5692 EN, The Netherlands

Abstract

ISA-95 was developed with the purpose of integrating enterprise and control systems. Enterprises use this standard to prepare and realize integration projects. It is less well known that the ISA models and terminology can also be used to analyze manufacturing companies. The Ordina ISA-95 and Manufacturing Execution System (MES) competence center developed an approach for the analysis of the "As-Is" and the "To-Be" situation, based on ISA-95. This approach has been applied for several clients with different objectives (e.g., Johnson & Johnson, AKZO Nobel, Philips, Abbott Laboratories). For example, one end user needed to list the current automation systems and platforms and compare the situations of several sites. Another had to develop a User Requirements Specification (URS) for an enterprise-wide tracking and tracing system (so comparison of the requirements of several plants was needed). A third company wanted to develop an MES automation strategy for the coming 5 years; here the ISA-95 models were used as a basis for an interview program. The ISA-95 analysis approach is suitable in all these situations and more.

Purposes of an ISA-95 Analysis

Suppose you want to list the automation systems that are used on your company's manufacturing sites. Or you're trying to reach agreement with colleagues from different departments about which responsibilities should be carried out where. Or you're making an overview of the requirements the company has for the new tracking and tracing system it's going to buy. Where do you start your study? How do you decide on the scope of your analysis? How do you know which questions to ask? How can you be sure that you are not forgetting anything important? How do you decide which people should be involved in the project? In all these situations and for all these questions, an ISA-95 analysis provides a solution. The ISA-95 models and concepts can be used as a basis for a very well-structured analysis. Each manufacturing company can have its own specific goal for such an analysis. In general, the purpose of an ISA-95 study is to describe the current operations situation, identify potential problem areas, suggest possible improvements, and detail the requirements for an organization's enterprise and control systems.

Preparations for the ISA-95 Analysis

Prior to the analysis, you should carry out several preparatory activities and make several decisions concerning the scope, content, methodology, and duration of the analysis. You will, for example, have to decide how you are going to collect the basic information for the analysis. The input for the analysis can be provided in many ways (e.g., documents, individual interviews with employees, group interviews).

Although the ISA-95 analysis devotes extensive attention to the scope and objectives of the study, you will need to form a global picture of them beforehand. Depending on the scope and the objectives, particular subjects will or will not be included, and input will be needed from specific employees or departments. The ISA-95 models can help to determine the scope.

If you have chosen the interview method, then decide who you're going to interview. When you determine the global scope, the subjects that should be covered become apparent. After this, it's time to determine which employees can provide input on these subjects.

After you've specified the scope and determined which employees are going to contribute to the interview sessions, it's time to schedule the steps of the ISA-95 analysis. As will become clear in this chapter, the steps are completely based on the ISA-95 models.

Deliverables of the ISA-95 Analysis

The analysis will ultimately deliver several tangible results. The first is a textual description of the current and the desired future situations. You could call this the blueprint. The purpose of the blueprint is to provide a textual explanation of the desired future situation and the source of the desires and requirements. Vendors of potential solutions can read this text to understand the background of the problem and to ensure that their solutions mesh as closely as possible with company-specific issues.

The desires and requirements in the blueprint can be repeated point by point in an appendix, giving rise to a checklist. The purpose of the checklist is to be able to determine the degree to which vendors' solutions fulfill the requirements and desires posited by the organization.

Finally, it is customary and also a good idea to explain the results, conclusions, and recommendations to management by means of a presentation. The management presentation forms the conclusion of the analysis phase. During this presentation, management is informed of the results of the analysis, and recommendations for the future are made.

Steps in the ISA-95 Analysis Information-gathering Phase

The ISA-95 analysis consists of the following steps:

- A tour of the company

- Determination of the final scope and the business drivers

- Interviews based on the functional hierarchy model (to get a first impression of the company, its departments, and its automation systems)

- Interviews based on the equipment hierarchy model (to obtain a description of the physical structure of the company and of its processes)

- Interviews based on the functional enterprise control model (to analyze the order processing function, the procurement function, the marketing and sales function, etc.)

- Interviews based on the Production Manufacturing Operations Management (MOM) activities model (to analyze the activities,

information flows, and automation systems within the production
department[s])

- Interviews based on the Maintenance MOM activities model (to analyze the activities, information flows, and automation systems within
  the maintenance department[s])

- Interviews based on the Quality MOM activities model (to analyze
  the activities, information flows, and automation systems within the
  lab[s])

- Interviews based on the Inventory MOM activities model (to analyze
  the activities, information flows, and automation systems within the
  warehouse[s])

- Interviews concerning other activities important on Level 3 (to
  assess the management of information, documents, and regulatory
  compliance)

Between and after the interview sessions you will need to write the blueprint
and list the requirements in the checklist. For each subject (e.g., order processing,
detailed production scheduling, or maintenance tracking), describe the current situation, the current automation systems, the issues and points of interest, and the
requirements for the future. Also, make a list of these requirements and give them
all a number. This will make communication with future vendors easier.

Describing the To-Be Situation

Describing the As-Is situation is the most important step, as this gives the complete
foundation for defining the To-Be target and also makes it possible to prioritize the
different paths to go from As-Is to To-Be. After describing the As-Is situation and
its issues and the requirements for the To-Be situation, the improvement possibilities need to be defined and prioritized in preparation for the business justification.
These improvements may be organizational improvements or improvements of
current systems, or they may be focused on the implementation of a new Level 3
automation system or the integration with other systems. If the implementation
of a new or improved MOM system is recommended, then the architecture of the
future MOM system can be defined based on the ISA-95 models. This system will
have to support the Level 3 production, maintenance, inventory, and quality activities for which requirements have been described during the ISA-95 study. For
some enterprises, scheduling may be complex, and implementation of a detailed

scheduling system may lead to many advantages for the enterprise. For other enterprises, scheduling may not be an issue at all, but quality data collection and tracking may be very important activities that can be optimized by implementing a new or improved system.

The Roadmap: How to Go from the As-Is to the To-Be Situation

Once you have defined the future architecture, the steps to realize this architecture have to be prioritized (Fig. 18.1). Often, the plan requires more than one project to implement parts of the complete architecture. A business justification and Return On Investment (ROI) study are often required for each project's funding. Priorities are defined based on the business drivers and management's expectations. The projects that bring the most advantages for the enterprise often have to be implemented first, but you may also want to start with a small pilot project (to test if the organization is ready for a new way of working.) Each project will have its own phases, like (if you follow the Good Automated Manufacturing Practice [GAMP] methodology) writing the URS; writing the Functional Specification (FS); writing the Design Specification (DS); building the system; and performing the Installation Qualification (IQ), Operational Qualification (OQ), and Performance Qualification (PQ).

In short, a roadmap can be developed by the following steps:

- Define future architecture

- Prioritize parts of the architecture

- Define projects for realization of the architecture in several phases, and prioritize these projects based on the priority of parts of the architecture

- Define project phases (based on a standard project approach like GAMP)

Best Practices

Anyone who's been "in the trenches" knows that things often go very differently in practice than as described in the standard. It's great to explain how an ISA-95 analysis can be carried out in an ideal situation, but in reality it isn't always that

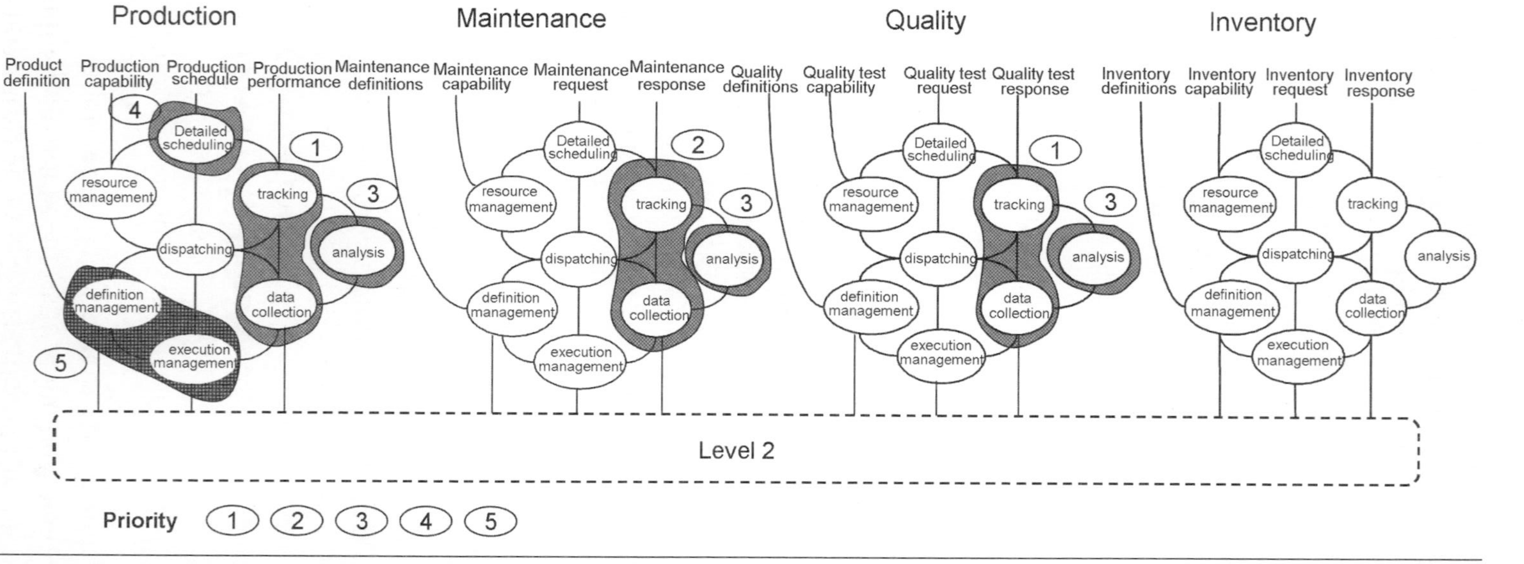

Figure 18.1. Example of a functional architecture defined after an ISA-95 analysis.

easy. Everyone who performs an ISA-95 analysis will encounter a few pitfalls and should be on guard for them. Here are some recommendations:

- Plan enough time to process the interview sessions.
- Estimate how much time the entire analysis will take, and try to complete it within that time frame.
- Give the participants enough time to cooperate with the analysis.
- Never skip the business drivers step. Make sure the right people provide input (e.g., the board of directors, management).
- Tell participants at the beginning of the interview sessions what you expect from them and what they can expect.
- Keep your eye constantly on the scope.
- Make sure that discussions don't take too long and don't get out of hand.
- Determine the right level of detail.
- Provide the text that participants should review in bite-size chunks.
- Describe the requirements point by point. Number each requirement. State just one requirement per number. A wish list can also be created.
- Ensure that the person in charge of the company has a final review of the requirements.
- After you've discussed the conclusions and recommendations with the client, invite all participants to a presentation in which they are again thanked for their efforts and can hear the conclusions and action items to which their efforts have led.
- Make regular backups of your documents, and save your changes regularly during interview sessions.

End User Quotes

In 2004, I performed ISA-95 analyses for the companies Aviko in Steenderen, the Netherlands, and Abbott Laboratories in Zwolle, the Netherlands.

Karst van der Pol, who is senior systems analyst at Abbott Laboratories in Zwolle, stated the following in an interview:

Last year, we started making an IT roadmap with our sister companies. Electronic batch records [EBR] were an important topic there. We want to achieve systems harmony. EBR was given the highest priority at that time. In particular, the enormous mountain of paper that all quality systems entail is a thorn in our plant manager's side. He's a member of the steering committee for the IT roadmap. All that paper is old-fashioned.

We had a communication problem with the sister companies, and also between the IT and QA [Quality Assurance] departments. We used different terminology, which makes it very difficult to achieve harmony. That's why we have to agree on standards and so we took a look at the available industrial standards. That's how we ended up at ISA-95. . . . If you asked us, "Would you do it again?" the answer is "yes." This is something you have to go through. We wrestled with the question, what do you put where? Now we're making an internal translation and we're investigating what we called something before, and what we should call it now using ISA-95 terms.

As head of the business desk, Meindert Visser has final responsibility for production and materials scheduling for Aviko B.V. and for the master data in SAP. In an interview, Visser said the following:

Our experience of the ISA-95 analysis in the form of a workshop was positive, mainly thanks to the exchange of perspectives from different disciplines aligned in one direction (the direction of the business goals). It creates solidarity and clarity, provides recognition and appreciation, and it helped to get everyone pointing in the same direction. It also helped us to form a crystal clear picture of something in a short time span. Based on the analysis, we concluded that we don't need a separate MES system, because extensive integration of the production (PP/PI), quality, sales, and distribution (logistics) modules from SAP with the production lines has already taken place. Also, questions about potential future situations can, to the extent we can foresee, be answered using the standard.

CHAPTER 19

Avoid a Mess by Building an ISA-95 Compliant MES

Presented at the WBF European Conference, November 13–15, 2006, by

Erna Noordkamp, Dr. Ir.
erna.noordkamp@ordina.nl
Ordina Technical Automation B.V.
Postbox 293
5600 AG Eindhoven, The Netherlands

Abstract

In the beginning of 2006, the ISA-95 and MES competence center of Ordina Technical Automation started to build a ISA-95 compliant Manufacturing Execution System (MES) framework. This application was built with Microsoft .NET techniques and is called Framework for MES Solutions (FMS).

The functionality of FMS focuses on the functions in the production and inventory operations management model that are not supported by other Enterprise Resource Planning (ERP) or MES software applications. Therefore, the main objectives of FMS are (1) the translation of a production schedule from Level 4 into orders and batches with product definitions or control recipes for production control in Level 2 and (2) the collection and management of information about the inventory and product genealogies.

The ISA-95 object models, the equipment hierarchy model, and the ISA-88 master recipe model and control recipe model were used as a basis to build FMS. The conclusion was that it is a large step from these models to a working software application framework. This is also a result of the fact that the ISA-95 object models are meant for information exchange between Level 3 and 4 and not for storage of information. Nevertheless, the object models form a strong and generic basis for FMS.

Introduction

We had several reasons for building our "own" MES framework. Among them is the reason that the existing MES and ERP applications do not have all the functionality that is necessary for production and inventory operations management in Level 3 of our customers' production plants. Another reason is that almost none of these applications are ISA-95 compliant, and we see high benefits from the standardization of information by the ISA-95 standard.

The functionality of FMS focuses on the functions in the production and inventory operations management models that are not supported by other applications. Figure 19.1 shows the generic activity model of Manufacturing Operations Management (MOM) and the functions that are supported by FMS, ERP applications, and other MES applications in general.

The functionality gap that is left by the existing ERP and MES applications is mainly the preparation and monitoring of production execution in Level 3 concerning the functions production dispatching, production execution management, and product definition management. The main objectives of FMS are the translation of a production schedule from Level 4 into orders and batches with product definitions or control recipes for production control in Level 2, and the collection and management of information about the inventory and product genealogies.

Framework for MES Solutions

Briefly, the functionality of FMS consists of the following:

- Equipment information management

- Materials information management, including inventory

- Information management of the equipment hierarchy and process segments

- Information management of product definitions and master recipes

- Management of production orders from Level 4 and the translation into batches with product definitions or control recipes

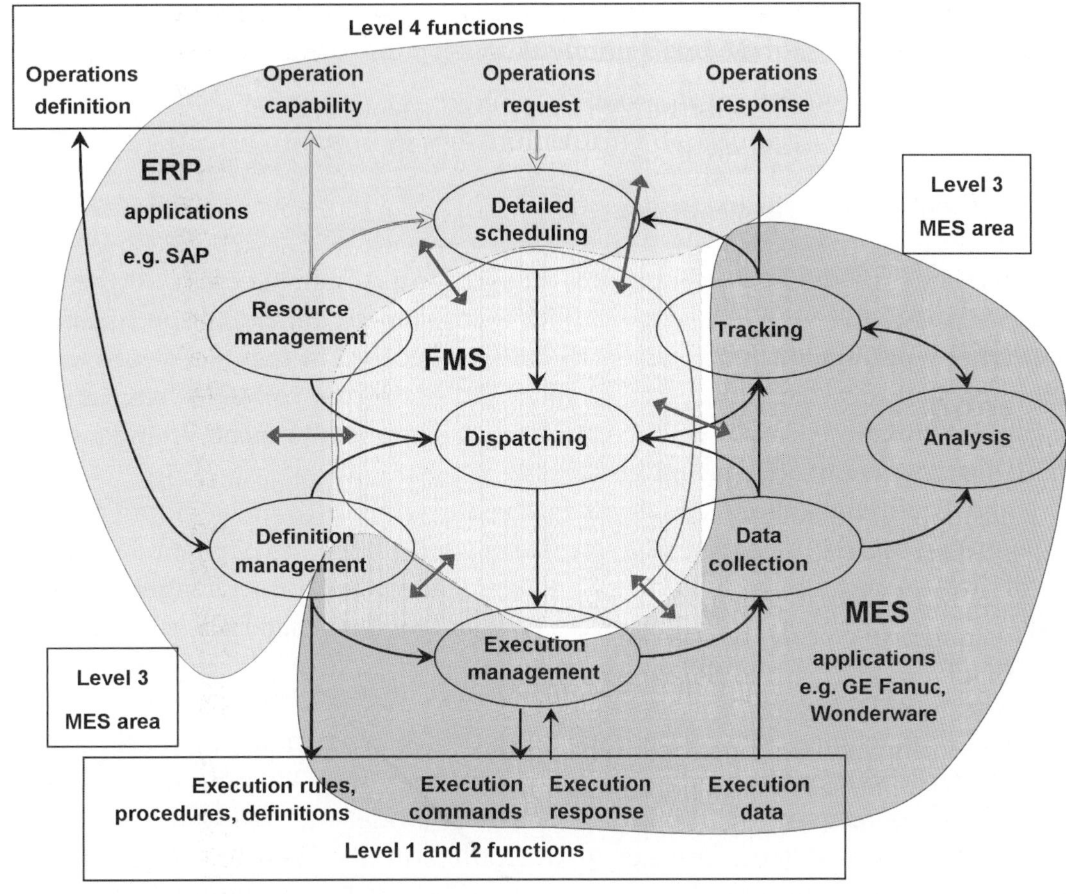

Figure 19.1. MOM generic activity model.

- Connection to an ISA-88 compliant batch application interface for downloading the batch information and control recipes for Level 2 and for collection of data from the batch application about the batches executed and materials produced and consumed

In addition to the functionality mentioned in this list, FMS contains functions typical for software applications (e.g., user management, security, and language settings).

FMS was developed based on several ISA-95 object models, the equipment hierarchy model, and the ISA-88 models for master recipe and control recipe. The following object models were partially incorporated into FMS:

- Equipment model
- Personnel model

- Material model

- Process segment model

- Product Definition model

- Production Schedule model

- Production Performance model

The Personnel model was only used as a basis for user management. At the time of this writing, it is not in the scope of FMS to implement personnel specifications, personnel requirements, and personnel actuals, because in our experience this is not important for customers. Also, Quality and Maintenance MOM functionality is not implemented in FMS. FMS has two main components: a server and a client built with Microsoft .NET techniques.

The server contains an SQL server database in which all data are stored, including master data and transactional data. The master data concern the modeling of a factory with respect to equipment (e.g., equipment model), process segments (e.g., equipment hierarchy model and process segment model), materials (e.g., material model), and product definitions or master recipes (e.g., Product Definition model and master recipe model).

The transactional data concern the execution of production processes. This is information about production requests (e.g., orders), with linked product definitions or control recipes (e.g., Production Schedule model and control recipe model) and the information about the executed production processes and batches (e.g., Production Performance model).

All data can be added, modified, or deleted by means of the FMS client. This client has a user interface containing seven main modules and about twenty-five screens.

The ISA-95 object models and the equipment hierarchy model were used as a basis to build FMS. These models are meant for information exchange between Level 3 and 4 and not for storage of information. Nevertheless, we decided to use the ISA-95 object models with the classes and attributes for data storage and for the class model of FMS. This is not necessary for ISA-95 compliancy, but it has the advantage that no internal conversion of information is necessary to generate ISA-95 compliant messages. The result is that the ISA-95 models can be clearly recognized in the structure and entity names in the database and class model of FMS, as well as in the user interface.

Furthermore, FMS was extended with the possibility to manage master recipes and control recipes. The ISA-88 master recipe model and control recipe model were used for this part of FMS. The end user has a choice to use only the ISA-95 part of FMS or extend it with the ISA-88 part. We incorporated the master and control recipe model in such a way that the ISA-95 part of FMS was not affected.

Another important standard we needed to take into consideration was the Business To Manufacturing Markup Language (B2MML) standard published by WBF. We want to be compliant with this standard too, because FMS will communicate with other applications that use this standard.

Some interesting issues concerning the ISA-95 object models, the ISA-88 recipe models, and the development of FMS are discussed in the following sections.

Material Management

The ISA-95 material model was used as a basis for the material management module of FMS. Figure 19.2 shows the part of the material model that is within the scope of FMS.

For the development of FMS, the ISA-95 definition of classes and attributes of the Material model was strictly followed. The Material model provides sufficient information to handle material management in FMS. This means, for instance, that only material sub-lots have an attribute "storage location." Therefore, only a material sub-lot can be found (physically) in a storage location (e.g., a place in a warehouse or a silo); no material lots will be found. A material sub-lot does not have properties, in contrast to material lot, material definition, and material class. Figure 19.3 shows the material definition screen of FMS.

The seven basic modules of FMS are presented in the lower left corner of the screen. The module Material Management is currently selected in the figure. This module contains five screens (menu in upper left) for management, respectively: Material Properties, Material Classes, Material Definitions, Material Lots, and

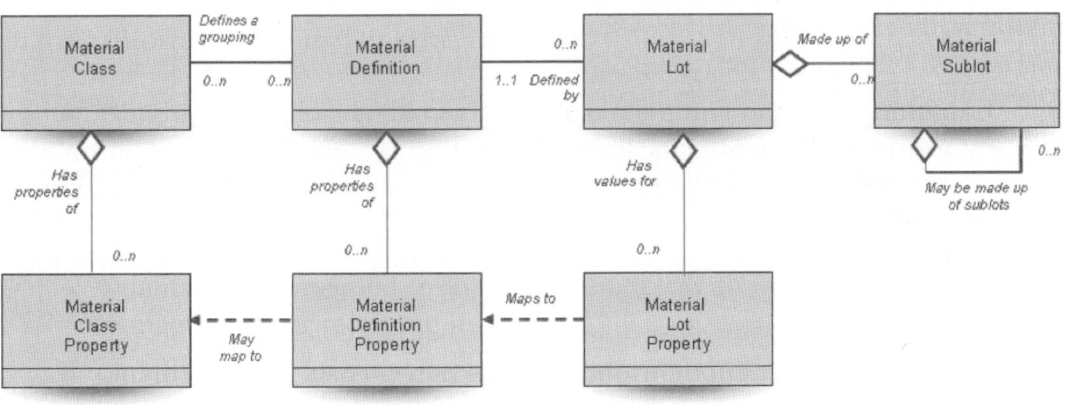

Figure 19.2. Material model.

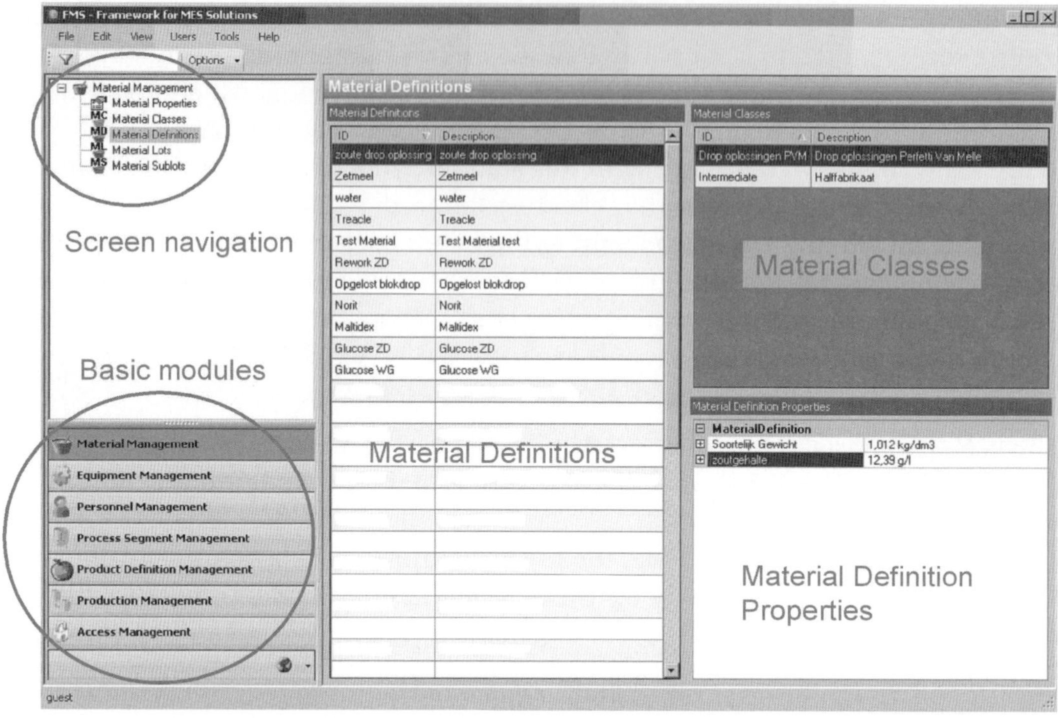

Figure 19.3. Material definition screen.

Material Sub-lots. The screen for management of Material Definitions is currently selected in the picture. In this screen, it is possible to add, modify, and delete Material Definitions instances; assign Material Definition instances to Material Class instances; and assign Material Property instances to Material Definition instances. The screen shown in Figure 19.3 is representative of most of the classes in FMS.

Equipment Hierarchy and Process Segments

The ISA-95 equipment hierarchy model was implemented in FMS because attributes of a process segment, material sub-lot, Production Schedule, and Production Performance message refer to the (storage) location and element type of this model. A location is an instance of an item from the equipment hierarchy model. Figure 19.4 shows that model.

In the ISA-95 standard, no attributes are defined for the items of the equipment hierarchy model. Therefore, we created our own location model that is similar to the ISA-95 equipment model and personnel model. The location model is shown in Figure 19.5. The class location has an attribute "element type." This is the name of an item from the equipment hierarchy. Figure 19.6 shows a picture of the locations screen in FMS.

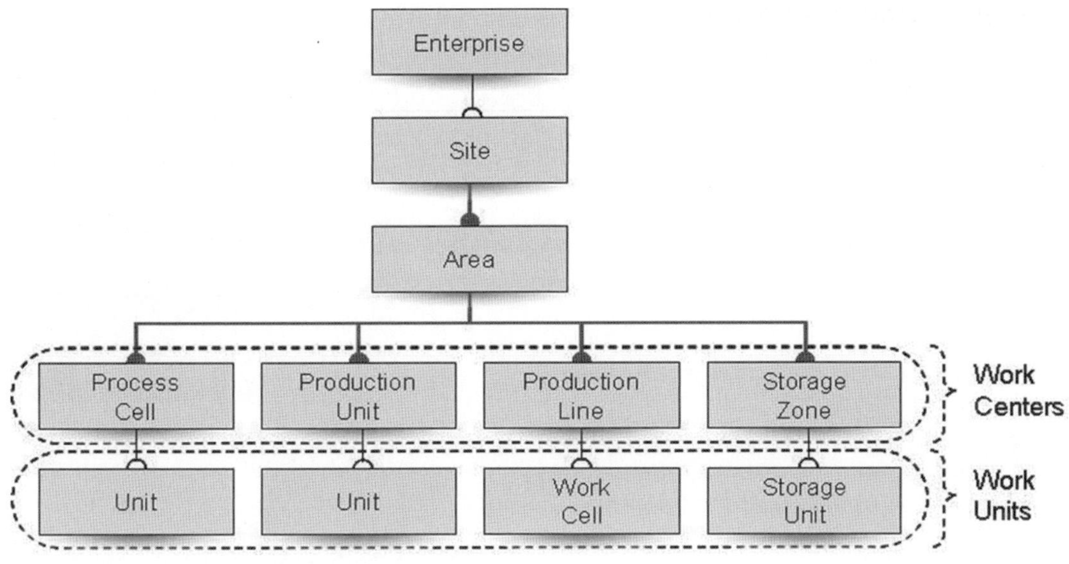

Figure 19.4. Equipment hierarchy model.

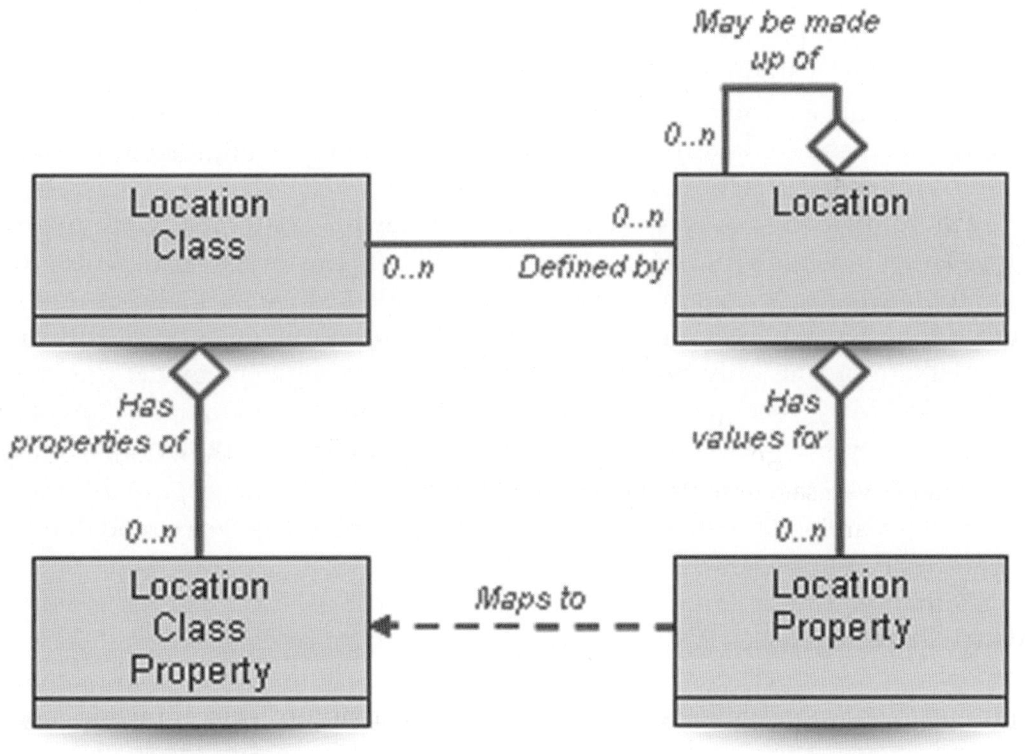

Figure 19.5. Location model.

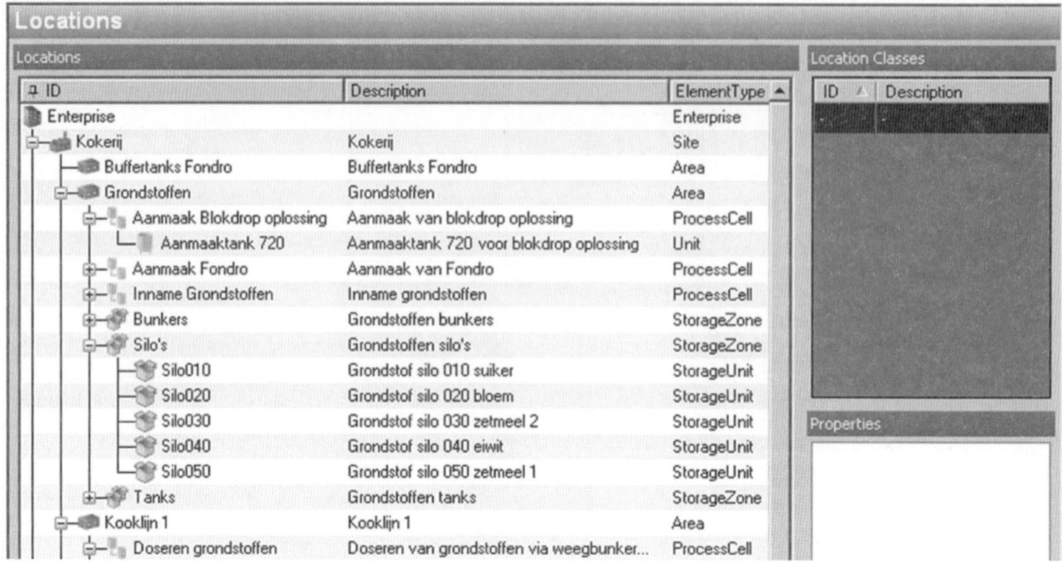

Figure 19.6. FMS Locations screen.

While modeling a factory in the locations screen, it is important to identify process items and parts of the factory and not merely the equipment in the factory. The equipment of the factory is modeled according to the equipment model in the FMS module equipment management. In FMS, it is possible to link a location instance to an equipment instance. In this way, the information about the equipment is available in the location.

According to the ISA-95 standard, materials can only be stored at storage locations. We have decided that storage locations can only be storage units. So any kind of buffer tank, silo, or place in a warehouse must be defined as a storage unit in a storage zone. No material can be stored in units of a process cell or production unit.

At the moment, only two classes of the process segment model are implemented in FMS: process segment and process segment parameter. Information about these two classes is needed for Product Definition, Production Schedule, and Production Performance. The other classes of the process segment model lie outside the scope of FMS at this moment because this information is not necessary to accomplish the main objectives of FMS.

Product Definition Management

Figure 19.7 shows an adjusted ISA-95 Product Definition model. In this model, only the classes of the Product Definition model that lie within the scope of FMS are shown, including the master recipe model of the ISA-88 standard.

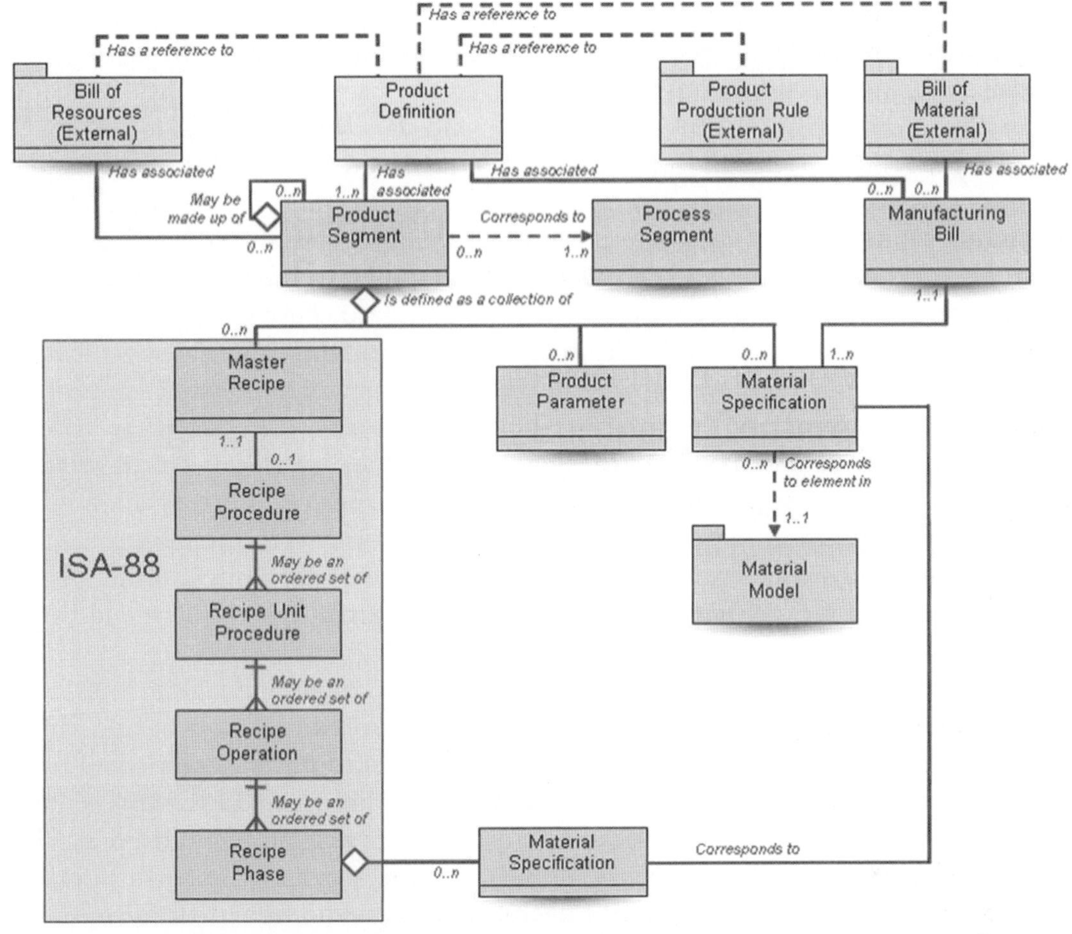

Figure 19.7. Adjusted ISA-95 Product Definition model.

The model in Figure 19.7 shows how product definition is implemented in FMS at the moment. For this model, we adopted a few definitions of the B2MML standard.

We renamed the ISA-95 class from "Product Production Rule" to "Product Definition" because this is the name used in the B2MML standard for this class. The attributes of the class "Product Definition" are similar to the attributes of the ISA-95 class "Product Production Rule." The Product Production Rule shown in Figure 19.7 is a class that exists in Level 4 and provides information for the Product Definition.

The equipment specification was deleted from the model because there is no need to store that information for a product segment at the moment. In the future, it may be added again when customers need this information in the ISA-95 part of FMS.

For the interface with an ISA-88 compliant batch application, we needed to define and store information about master recipes. After a lot of discussion, the model in Figure 19.7 was constructed. It gives the user the possibility to define products according to the ISA-95 definition and as master recipes according to ISA-88. The ISA-88 part is optional and does not need to be used by a customer.

A master recipe is part of a product segment. If a master recipe is defined for a product segment, then the product segment must correspond to a process segment for a process cell because a master recipe is always created for processing in a process cell. This is according the ISA-88 standard. Furthermore, a material specification must be made in the Product Segment, and this material specification must correspond to the material specification(s) of the recipe phases (i.e., process inputs and process outputs).

It is up to the user to use either master recipes or the material specifications and product parameters for product definition. In a master recipe, more detailed information can be specified, especially about the units and the production procedure. On the other hand, a master recipe is created for a process cell, while product definitions and product segments can be created for all hierarchical levels in the factory.

Production Management

The Production Management module of FMS is a combination of the ISA-95 Production Schedule model and Production Performance model. The reason for this is that the structure of the two class models and the attributes of corresponding classes are the same. Only the names of the classes are different. Furthermore, in this way data storage in the database is more effective, and in practice, identifications (IDs) of instances are often the same.

For example, the ID of a production request is often the number of a production order that is generated in Level 4 by an ERP application. The ID of the production response that corresponds to the production request must then be the same because the response information is returned to the ERP application.

The models are put together by adding the attributes of corresponding classes of the ISA-95 Production Schedule model and Production Performance model to new classes. In two cases, we gave the resulting class a new name. The classes "production request" and "production response" were added together into a new class "production order" that has the attributes of both classes. A production order is made up of production order segments that contain the attributes and classes of "Segment Requirement" and "Segment Response."

In the user interface of FMS, a screen is available to add and modify production orders. This screen shows the request information and the available response information of a production order. It is often desired by customers to see both target values (i.e., request information) and actual values (i.e., response information) at the same time.

Finally, it is possible to create a production order segment with a control recipe. In this case, the production order segment is linked to a master recipe, and this master recipe is copied to a control recipe. In FMS, it is possible to create production order segments with information structured according to ISA-95 or ISA-88.

Figure 19.8 shows the adjusted Production Schedule model in the way it is implemented in FMS. We call it the production order model.

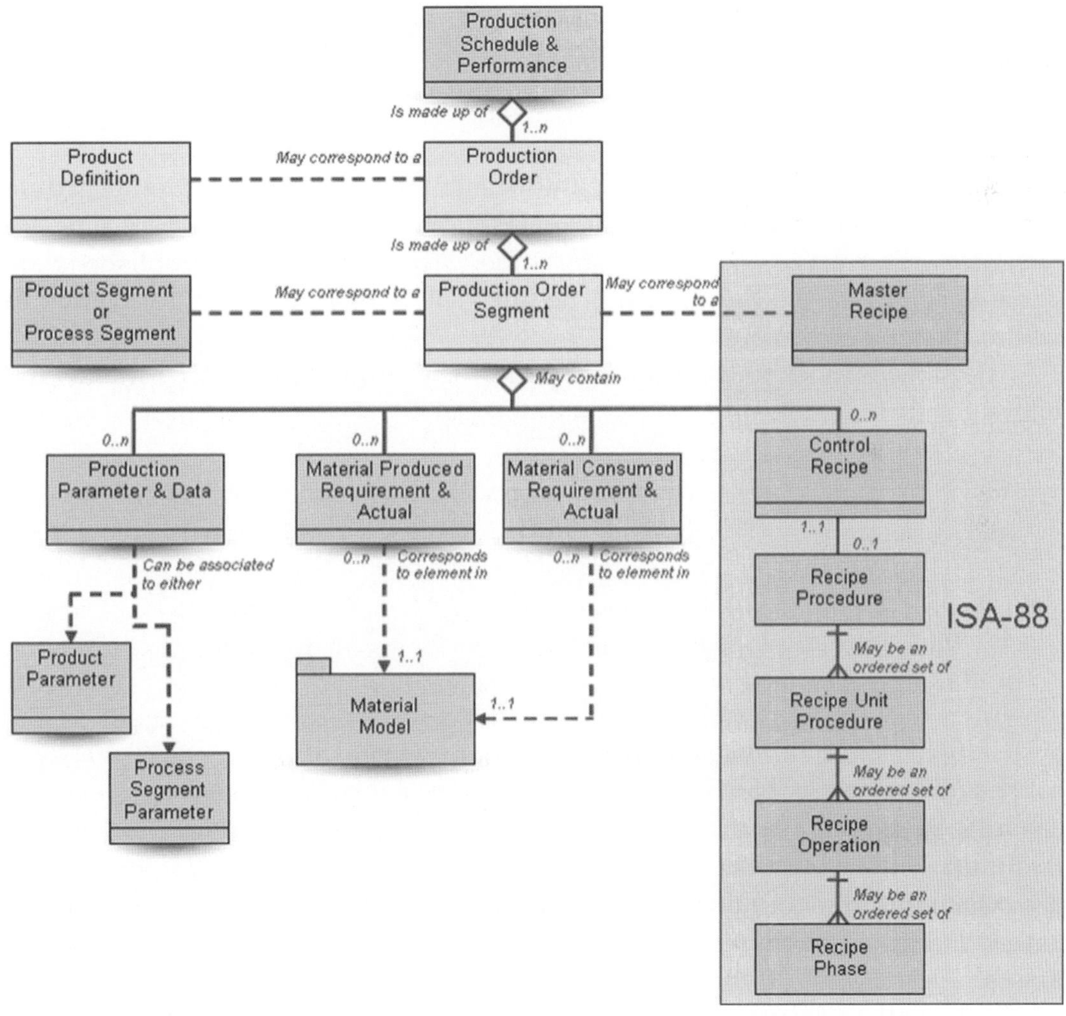

Figure 19.8. Adjusted Production Schedule model.

The Production Schedule and Performance class has the following attributes:

- ID
- Description
- Target start time
- Target end time
- Actual start time
- Actual end time
- Published date schedule
- Published date performance
- Location
- Element type

The result is one class containing the information about the Production Schedule and Production Performance. The ID, description, location, and element type are the same for both classes. The target start time, target end time, and published date schedule belong to the Production Schedule. The remaining attributes belong to the Production Performance.

The same applies to a production order. The production order contains the following attributes:

- ID
- Description
- Product definition
- Target start time
- Target end time
- Actual start time
- Actual end time
- Priority

A production order contains the information from a production request and the corresponding production response. The ID and description are the same for both classes.

For the other classes in the production order model, the attributes have been put together in a similar way. In FMS, extra attributes have been added to classes in the models for different reasons. The main reasons are that extra information is desired by customers (e.g., a production order status) or that extra information is necessary for the execution of FMS.

As shown in Figure 19.8, the personnel requirement, equipment requirement, and consumable expected of the Production Schedule model classes and the corresponding classes of the Production Performance model (i.e., personnel actual, equipment actual, and consumable actual) are not implemented in FMS. They are out of FMS scope at the moment.

The ISA-95 part of a production order segment may contain production parameters and data, material produced requirements and actuals, and material consumed requirements and actuals. If the ISA-88 part of the model is desired by a user, then the production order segment must be linked to a master recipe, and a material produced requirement must be created in this segment. Thereafter, batches with control recipes are created automatically based on the default batch size defined for the master recipe and the quantity defined for the material produced requirement. If the production processes in Level 2 are controlled by an ISA-88 compliant batch application, then it is necessary to create batches with control recipes in the production order segment.

Conclusion

It is a large step from the ISA-95 object models and the equipment hierarchy model to a working software application framework. This is a result of the fact that the ISA-95 object models are intended for information exchange between Levels 3 and 4 and not intended for storage of information. Nevertheless, the object models form a strong and generic basis for FMS. This is one of the advantages of using the ISA-95 models as a backbone for your application—the models and the information that are exchanged are discussed worldwide by experienced ISA members, and they are widely accepted as the best solution.

Another advantage is the use of ISA-95 terms and models in the FMS user interface. It is relatively easy for users that are familiar with the ISA-95 models and terminology to operate the screens of FMS and to insert the correct information. It is somewhat easy for them to understand the functioning of FMS.

In FMS, most of the ISA-95 object models are only partly implemented. In each object model, classes are defined that are not in the scope of FMS at the moment. The main objective of FMS is to provide basic Level 3 functionality that is not provided by other MES or ERP software applications. Additionally, maintenance and quality test information does not belong to the scope of FMS. This chapter describes the classes of the object models that were used to build FMS.

These classes were used to build a class model for FMS and a data model for the database. All attributes of the ISA-95 classes used were incorporated into FMS. Sometimes attributes or other information are added to the classes for different reasons. The principle reasons are due to extra information that is desired by customers or for the execution of FMS.

In a few cases, we added a class model or made adjustments to the ISA-95 object models. For example, we added the location model based on the equipment hierarchy model and joined the Production Schedule model and Production Performance model to create the production order model.

Furthermore, FMS was extended with the possibility to manage master recipes and control recipes based on the ISA-88 master recipe model and control recipe model. The end user has a choice to use only the ISA-95 part of FMS or extend it with the ISA-88 part. Management of master and control recipes is necessary if the production processes in Level 2 are controlled by an ISA-88 compliant batch application that must be controlled by FMS. We incorporated the master and control recipe model in such a way that the ISA-95 part of FMS was not affected.

Building MES Applications with ISA-95

Presented at the WBF European Conference, October 11–13, 2004, by

Ing. Erwin Winkel MBA

erwin_winkel@bat.com
BAT Manufacturing BV
Kerkstaat 27
6901 AA Zevenaar, The Netherlands

Abstract

In 2002, British American Tobacco (BAT) Manufacturing developed Manufacturing Execution System (MES) applications around an Oracle Relational Database Management System (RDMS). The MES applications are used to provide operators with near real-time information about their production process (i.e., the making and packing of cigarettes). The database model is partially based on the ISA-95 object model. For the interfaces to other information systems, the ISA-95 interface definitions are used. The messages between systems are based on Extensible Markup Language (XML) and the WBF XML schemas (as much as possible). The technical architecture is based on the Oracle Application Server and Java using the Java 2 Platform, Enterprise Edition (J2EE) guidelines. The user interface is completely Web based. The modules that have been implemented at this point are the following:

- *Material control*. Manually scanning each material with a barcode scanner before the product is used on the machine

- *Quality control*. Automatically capturing the test data and displaying at the machine with appliance of Statistical Process Control (SPC) rules

- *Planning control*. Automatically capturing real-time production data and verification against planning and shift targets

More modules are planned. For analysis and reporting, a manufacturing data warehouse module has been built, again using an Oracle database. Reporting is done with business objects.

Introduction

In modern manufacturing processes, the usage of automated systems is inevitable. Whether for production control, scheduling, or reporting, computers are used. One of the objectives of MES systems is to capture data out of the production process and transform these data into valuable information. This chapter describes the recent development and introduction of several MES modules in a manufacturing plant. The role of the ISA-95 standard in the project will be highlighted. After a short introduction of the production process, the business goals for the project will be briefly discussed. After this, the functional and technical architecture will be presented. The results and benefits of the project will be discussed as well as the benefits of using the ISA-95 standards during the project. The chapter ends with an explanation of the lessons learned.

Problem Description

Providing information on a shop floor can involve several problems: (1) the information needs to be timely in order to let the operators be able to adjust the process; (2) it needs to be accurate because decisions about the process need to be taken immediately; and (3) it needs to be presented in an operator-friendly way. These requirements place strong demands on the IT systems in use. When systems are tied to the process itself, this is further complicated by the demand for availability. Because of the variety in manufacturing processes, it is extremely difficult to implement commercial off-the-shelf packages. The solution described in this chapter is to develop an MES environment that makes use of standard models and concepts. In this way, flexibility and extendibility, as well as easier maintenance of the system, are built in.

Project Context

The MES system described in this chapter was built at a cigarette manufacturer. The factory produces approximately twenty-seven billion cigarettes per year. The production process basically consists of two steps:

1. Processing the raw tobacco leaves and blending

2. Making and packing the cigarettes

Step 1 is a typical batch-oriented process. The blend mixtures consist of 8000 kg batches that are processed in approximately 10 hours processing time. This process is done in the Primary Manufacturing Department (PMD).

After the tobacco blends are ready, the tobacco is used for making and packing the cigarettes in step 2. The making and packing is done in the Secondary Manufacturing Department (SMD). This is a typical discrete manufacturing process. Production runs are scheduled at production lines, based on customer orders or existing stock level. Currently, BAT Manufacturing has forty making and packing combinations to produce production runs. The MES system that is discussed in this chapter is situated in the SMD and is referred to as "SecIS" (Secondary Information System) throughout this chapter.

Project Approach

MES systems typically are situated between the Enterprise Resource Planning (ERP) systems and the Production Control Systems (PCS). Figure 20.1 provides a visual representation of this. One of the challenges in an MES environment is to build the interfaces required to exchange data between the different layers.

The SecIS project started in 2000 with the drawing up of an inventory of the existing solutions available at other BAT Manufacturing plants in Europe. Since these solutions were all tailored to the local situation, they could not be easily reused at a location in Zevenaar. The decision was then made to start with a new implementation in the Zevenaar plant.

First, the functional requirements for the MES system were defined. This was done by making use of the ISA-95 standard. The objects from parts 1 and 2 of the standard were used as references to define the functional requirements.[1]

Based on the functional requirements, a short list of the existing commercially available packages was investigated. This was done in 2001. At that time, the commercial packages did not fit the requirements, and therefore a decision was made

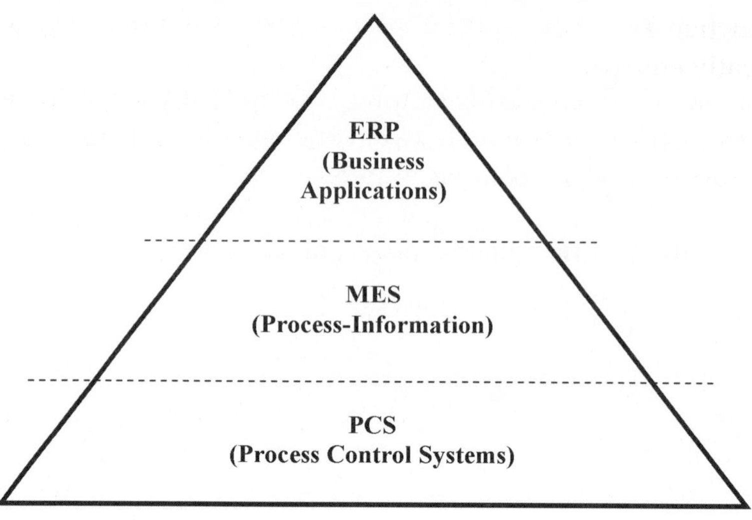

Figure 20.1. Automation pyramid in a manufacturing environment.

to build a new MES environment (SecIS) at the location in Zevenaar. The following functional modules were defined:

- *Performance control.* This module captures machine data directly at the production machines. Data on downtime, performance, yield, and so on would become available for on-line reporting and feedback to the operator.

- *Quality control.* This module captures the quality measurements data. The data can then be used for SPC techniques as well as problem and trend analysis.

- *Planning control.* This module brings the scheduled production orders on-line and matches actual production against it. At any moment in time, the state of the order can be examined.

- *Material control.* All materials are applied with a barcode containing the article number. When they are used at the machine, they are scanned by the operator and verified against the manufacturing bill. Wrong materials are identified, and the operator is warned that something is wrong. This module contributes to the "first time right" principle.

- *Waste control.* Waste is created during the process. The target is to eliminate as much waste as possible. By measuring the waste per production line and reporting on it, on-line visibility for the operator is greatly enhanced.

After the functional requirements were defined, the technical architecture was chosen. The current existing technical architecture and strategic corporate guidelines were taken into account when making choices about the technical architecture for the SecIS environment. The current environment is based on the following:

- Web technology
- Oracle database and Java (J2EE)
- A message queuing communication infrastructure (WebsphereMQ)
- Open standards (as much as possible)

The following sections list business goals that were defined for the project.

Empowerment of the Shop Floor Level

In order to empower the operators and allow them to be able to make decisions, they need information. Therefore, SecIS is a state-of-the-art information system primarily targeted at the operators.

Development of Operators

"Machine operators" should develop into "process operators." For this to happen, the operators need to have information about the process as a whole, rather than only the condition of their "own" machine.

Efficiency and Effectiveness Improvements

Of course, the project did define some quantitative improvements as well. Improvements in the quality of the product as well as the performance of the machines were specified in the list of business goals. SecIS provides near real-time information on output, quality, and performance parameters. This yields the following improvements:

- Less "nonconformity" to the manufacturing bill and therefore less faulty products
- SPC and therefore a more accurate product

- More precise production in accordance with the schedule and therefore less waste

- The elimination of error-prone, administrative work at the machine

Preparation for a Faster-changing Environment that Becomes More and More Complex

The environment becomes ever more complex. Various versions of the end product lead to smaller runs. Brand changes increase the risk that something can go wrong. Adequate information systems support the operator to produce good quality products within reasonable time.

Where Are We Now?

To date, three modules are operational in the SMD:

- Material control

- Quality control

- Planning control

SecIS has contributed to the following production improvements:

- Empowerment of the shop floor level
  - Proof that materials are sometimes not within spec
  - Proof that the scheduling process needs improvement
- Development of operators and efficiency improvements
  - Improved ways of working and procedures.
  - Operators starting to take responsibility for the whole process
- Preparation for a faster-changing environment that becomes more and more complex:
  - An MES environment that is extremely flexible and allows us to deal with almost any situations

The results on effectiveness and efficiency are as follows:[2]

- Quality improvements

- ❑ Additional and timely samples

    - ❑ SPC techniques implemented

    - ❑ An increased Manufacturing Quality Index (MQI),[3] from 81%
      (January 2004) to 83% (August 2004)—a trend that is increasing

- ■ Productivity improvements

    - ❑ Less rework

    - ❑ A reduced number of Bill Of Materials (BOM) incidents, from a
      monthly average of eight (January 2004) to a monthly average
      of four (August 2004)—a trend that is decreasing

- ■ Waste reductions

    - ❑ Better conformance to targets

    - ❑ Awareness

    - ❑ A reduced number of rejects, from 2% (January 2004) to 1%
      (August 2004)—a trend that is decreasing

Summary and Lessons Learned

MES systems improve the quality of information needed at the shop floor level in
several ways:

- ■ Information becomes available in a timely fashion. This make adequate reactions from the operators possible.

- ■ Information is reliable. By capturing the data at the source, the reliability of the information is greatly enhanced.

- ■ The process becomes transparent. By providing information about
  the whole process rather than a single machine, it becomes clear how
  the process is running. This prevents suboptimization and enhances
  responsibility. It also allows bottlenecks to show up more clearly
  when the process becomes transparent.

The overall advantage is that the process becomes more stable and output is more
predictable.

The following sections list the lessons that were learned from the SecIS project.

MES Projects Need Long Decision Tracks

Circumstances are subject to change over the course of a project. During the SecIS project, a change in the management team occurred that led to a redefinition of the project. Other changes to take into account are the following:

- Production circumstances (e.g., new machines that need to be connected)
- Revised products or manufacturing bills that differ from anything that currently exists
- Shifts in focus (e.g., on quality instead of performance)
- Revised requirements due to external factors (e.g., global guidelines, new environmental rules)

MES Still Needs Complex Technical Architectures

Although interfaces and software integration are further developed, it is still extremely complex to integrate all components into one system. This is often because production systems tend to be quite old and do not always have the latest technological developments. This means that the following is true for an MES project:

- Many different skills are needed
- Strong project management is needed
- A lot of custom building is still necessary
- You have to connect *all* of it

Standards Do Help

Standards that were used in this project are ISA-95, J2EE, and EAN128. They were used as

- a reference when defining the requirements;
- object models and their attributes;
- templates during development (e.g., WBF XML schemas).

Since this was the first ISA-95 project for both BAT Manufacturing and the system integrator (ICT Solutions), clearly there were some learning effects. Not all choices made in the beginning were the best possible. Therefore, in 2004, some restructuring of the applications and design took place to include new insights.

The use of standards greatly enhances the communication between the different parties involved. Also, communication between different systems is made easier when definitions are uniquely defined and clear to everyone.

Experiences from the system integrator have clearly shown that the use of the ISA-95 object models and database schemas have contributed to a faster development process. The development of an MES framework based on ISA-95 will greatly enhance future developments in the MES environment.

Beware of Exceptions (Everywhere!)

In a production environment, there are always exceptions that need to be taken into account:

- Special (old) machines
- Special manufacturing bills
- Special procedures
- Special materials

Having an MES environment that is flexible enough to cope with these exceptions is an enormous advantage. Until now, we did not encounter a situation that we were not able to tackle.

Existing Culture Is as Important as Technology

In a production environment, procedures and work instructions can be quite old. Old habits are difficult to weed out. People need time to learn a new system, and they need to learn to trust the system. This means that a lot of attention is needed during the introductory phase. During the SecIS project, a lot of effort was put into training the operators (about one hundred and fifty persons), quality specialists (twenty-five persons), and operational management (twelve persons). Even then, trust is difficult to obtain, and the system has to prove itself over a long period before people believe what the screen shows them.

Notes

1. At that point in time, ISA-95 parts 1 and 2 were still being drafted. Part 3 of the standard did not exist. Now, part 3 would be used for this step.
2. All tangible improvements mentioned are based on the period from January 2004 through August 2004.
3. This is a formula to express the overall quality of the SMD.

Index

Consumer Packaged Goods (CPG),
213–14
context aggregation, 120
continuous process workflow, 125–26
control systems
    data translation for business
      decisionmaking, 127–40
    DDTM methodology, 107–13
    low-risk integration in, 140
    MES implementation and process
      analysis, 179
    production state definition, 134–36
    Unified Production Model (UPM), 24
Coordinated Universal Time (UTC),
151
cost analysis
    MES implementation, 174–75
    systems integration and, 144
cultural issues, 267
current system analysis, 178–80
customer value stream, 19–20

data acquisition systems, 179, 182–84
data analysis
    B2MML-XML data format
      integration, 93–98
    control systems data translation for
      business decisionmaking, 127–40
    flexible discrete manufacturing,
      53–55
    historical data, 130–33
    ISA-95-SAP R/3 integration, 158–59,
      158, 163
    lot genealogy, 137–38
    manual data collection, 130
    operations benefits, 136
    real-time actual operations
      reporting, 136–37
    Unified Production Model, 30
data coupling, 97

decision tasks
    data translation for business
      decisionmaking, 127–40
    ISA-95 MES applications
      construction, 266
Demand-Driven Supply Networks
(DDSNs)
    evolution of, 3–4
    interactive models, 5–8
    MAF and Lean manufacturing
      revival, 19–20
Directed Data Transaction Model
(DDTM)
    batch data transactions, 100
    "middleman" elimination, 105–13
Discrepancy and Corrective and
  Preventive Action (CAPA) tracking
    model, 39–41
    plant model implementation, 32
discrete manufacturing
    ISA-95 flexibility and, 45–63
    nontraditional operations, 46
    research background, 46–48
    workflow, B2MML best practices
      applications, 125–26
Discrete To Business (D2B) Working
  Group, 172
Distributed Control System (DCS)
    data transactions and, 104–5
    DDTM methodology and, 107–13
downtime operational state, 134–36

efficiency improvements, 263–64
engineer to order integration, 120
Enterprise Core Component (ECC), 90
enterprise platforms, 209–11
Enterprise Resource Planning (ERP)
    B2MML applications, 141–51
    best practices applications, 116–23
    business system implementations, 129